power mechanics

First Edition

Lithographed in U.S.A.

DR. WILBUR R. MILLER
University of Missouri — Columbia
Columbia, Missouri

DR. RALPH C. BOHN
San Jose State College
San Jose, California

ANGUS J. MacDONALD
San Jose State College
San Jose, California

DR. MARION E. MADDOX
University of Arkansas
Fayetteville, Arkansas

LAVON B. SMITH
Fayetteville, Arkansas

basic industrial arts

McKNIGHT
Publishing Company
Bloomington, Illinois

Library of Congress
Card Catalog Number: 78-53394

SBN: 87345-790-0 Paperback
SBN: 87345-798-6 Hardbound

TABLE OF CONTENTS

Chapter 1 OVERVIEW OF POWER AND CAREERS

Technology has developed because people have learned to harness and use power beyond their muscles. In early efforts, they used the natural power they found in the rivers, wind, and sun. Since then, people have developed power sources that are dependable. They have put to use different kinds of energy. These include mechanical, nuclear, heat, light, chemical, and electrical energies. These sources of power have played a vital role in the progress of technology. Electrical power has become a key factor in industrial production.

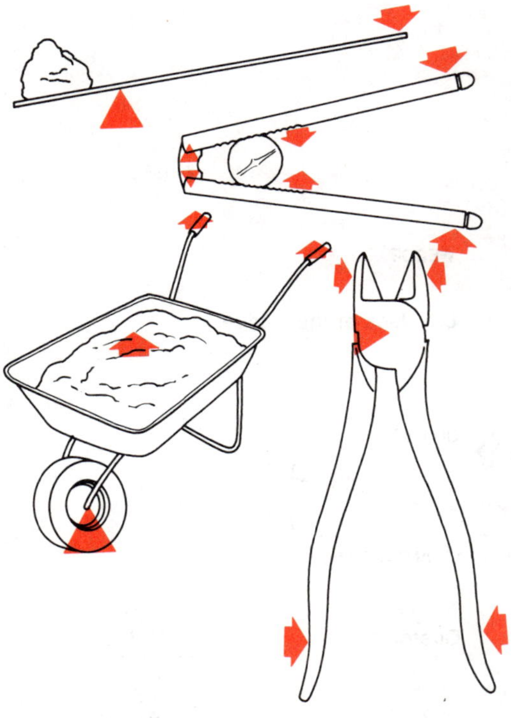

Fig. 1-1. Uses of the lever.

An energy crisis threatens the industrial world. It has become everyone's problem. Energy must be conserved. New forms of energy must be harnessed to meet the growing needs for power.

SIMPLE MACHINES

The only power that primitive people had was their own muscles. Later, they used the power of domesticated animals. But muscle power alone did not meet people's needs. In time, people developed six basic machines to perform work. These simple machines are the lever, pulley, wheel and axle, inclined plane, screw, and wedge. See Figs. 1-1 through 1-6. All machines use some form of these six machines. The pulley and the wheel and axle are basically levers. The wedge and screw are planes that have been modified or changed. Many complex machines are combinations of simple machines.

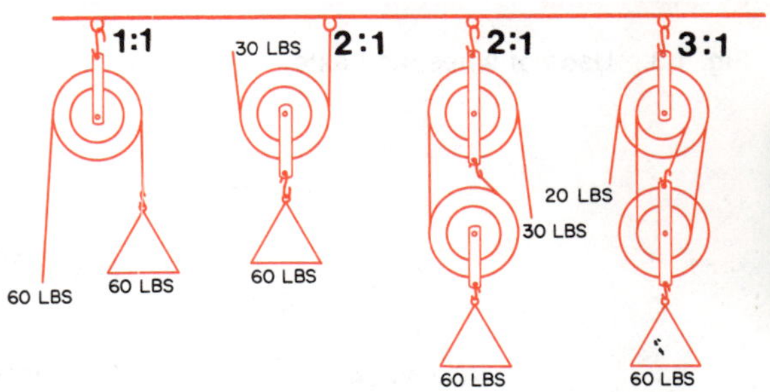

U.S. Customary	SI Metric
30 lbs.	13.6 kg
60 lbs.	27 kg

Fig. 1-2. Use of pulleys to gain mechanical advantage.

SI METRIC MEASUREMENTS

Since this book is an early experience in power, English terms of measurements are used. This will help you understand power and the ways it is used. Metric power measurements use the newton (N) for force, the joule (J) for work, the watt (W) for power, and the pascal (Pa) for pressure. You may be more familiar with common terms such as meters and kilograms.

Fig. 1-3. Uses of wheel and axle.

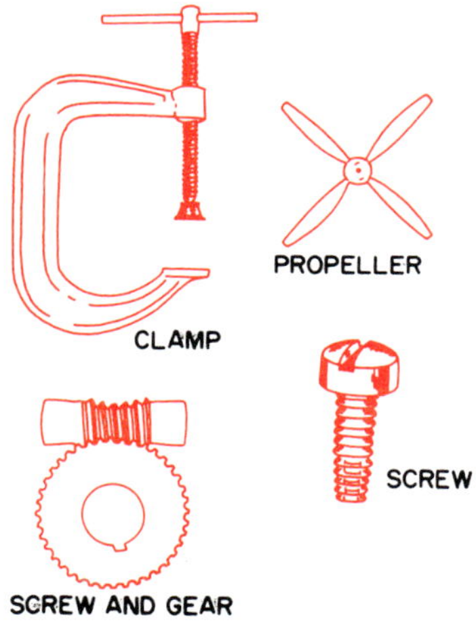

Fig. 1-5. Uses of the screw.

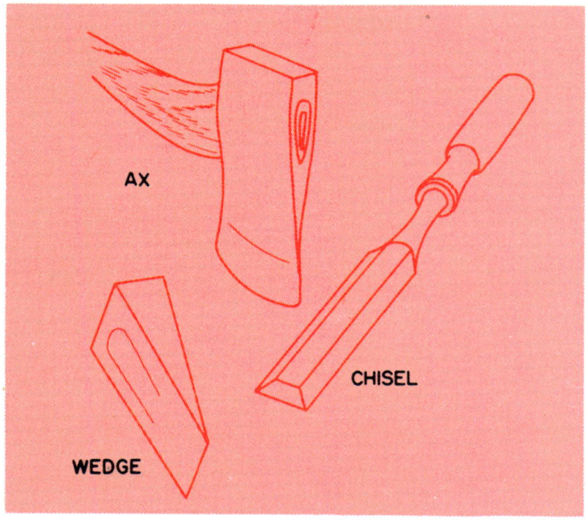

Fig. 1-6. Uses of the wedge.

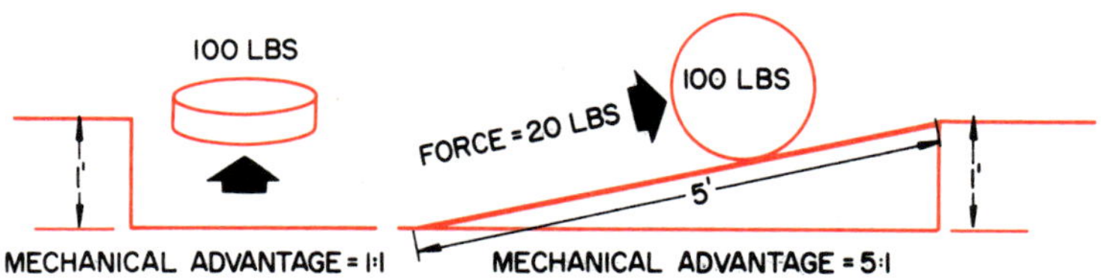

Fig. 1-4. Example uses of the inclined plane.

U.S. Customary	SI Metric
1 ft.	0.3 m
5 ft.	1.5 m
20 lbs.	9.0 kg
100 lbs.	45.4 kg

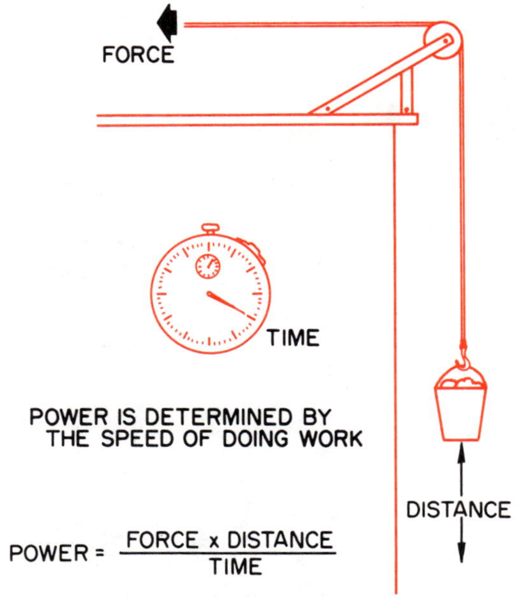

Fig. 1-7. Power is measured by the rate or speed of doing work.

For the most part, metric values will be shown as footnotes. These metric conversions will introduce you to the terms and values for future use.

POWER

Before you can understand simple machines, you need to know these terms:

1. **Power**—the time rate of doing work.
2. **Energy**—the capacity to do work.
3. **Force**—a push or pull.

Figure 1-7 shows that power is measured by the speed of doing work. Suppose one horse requires one minute to raise a 165-pound weight to a height of 50 feet. Then two horses could raise the same weight the same distance in one-half the time. The amount of work done in each case would be the same. It is 165 pounds times 50 feet equals 8250 foot-pounds.[1] (Note the metric conversion below.)

SI Metric Conversion

1. A 75 kilogram weight produces 735 newtons (N) of force. Over a distance of 15 meters (m), this force produces 11 025 joules (J).

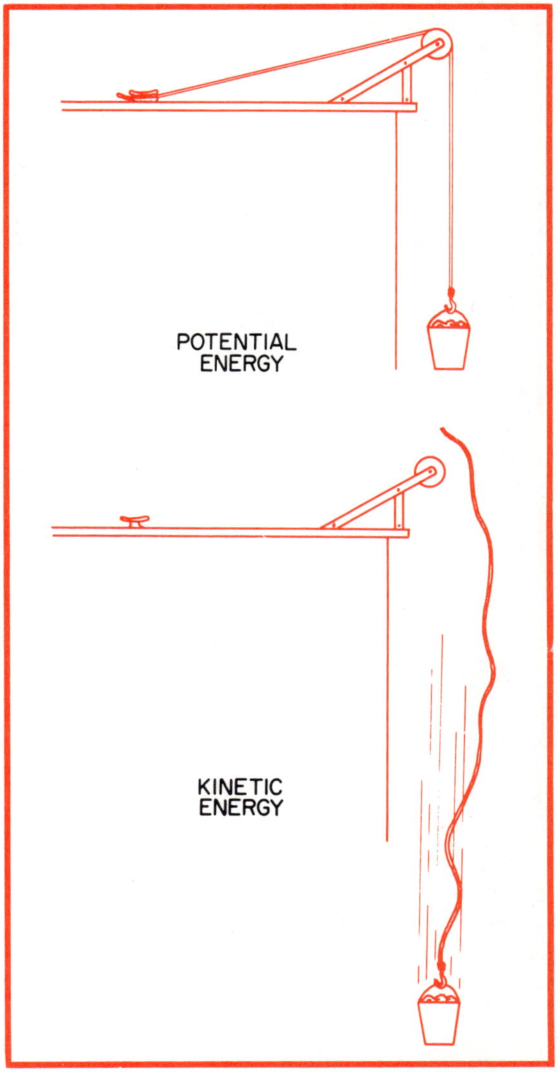

Fig. 1-8. Suspended weight is potential or stored energy. It is inactive until there is motion. Kinetic energy is energy in motion. The change takes place when potential energy is released from suspension.

Whether it takes one-half minute, five minutes, or longer, the work done is the same. The difference is in the amount of time required to do the work. To determine power, you need to know how long it takes to do a certain amount of work.

Fig. 1-9. Force may start movement of an object that is at rest. It may also change the movement of an object that is in motion.

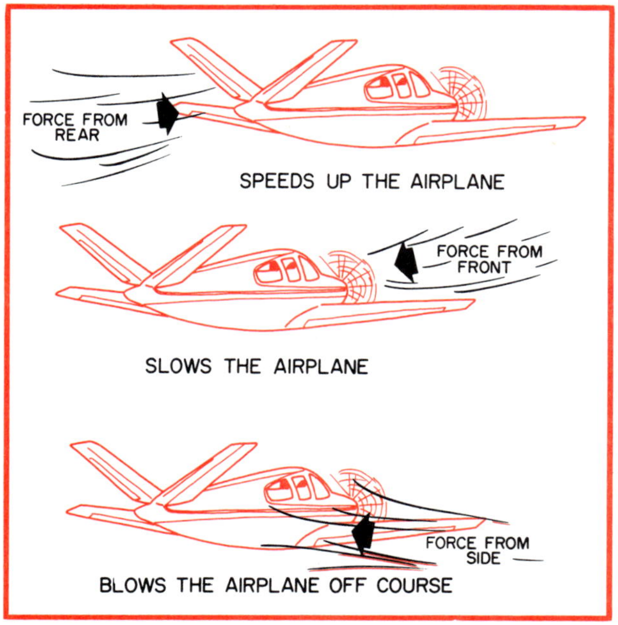

Fig. 1-10. The wind is a force. In aviation terms, it may be described as a tail wind, a head wind, or a cross wind.

$$\text{Power} = \frac{\text{Work}}{\text{Time}}$$

or

$$\frac{\text{Force (pounds)} \times \text{Distance (feet)}^2}{\text{Time (minutes)}}$$

Energy is the capacity or ability to do work. There are two kinds of energy. They are **potential energy** and **kinetic energy.** Study the examples of these two kinds of energy in Fig. 1-8.

Force is involved when energy is used to do any kind of work. In Fig. 1-9A, movement is started by the force of the club hitting a ball which is not moving. In Fig. 1-9B, force is shown as an action that changes the movement of an object already in motion. For example, the direction that a ball is moving is changed when it is hit.

SI Metric Conversion

2. This formula converts to:

$$\text{Power} = \frac{\text{Force (newtons)} \times \text{Distance (meters)}}{\text{Time (seconds)}}$$

This same principle applies to an airplane in flight. In Fig. 1-10A, the wind is a force behind the airplane that helps it fly faster. In Fig. 1-10B, the wind is a force against the front of the plane. It causes the plane to slow down. In Fig. 1-10C, the wind is blowing from the side of the plane. It causes the plane to fly in the wrong direction unless the pilot makes some adjustment to offset this.

MACHINES

Machines are used in these ways:

1. **To convert one form of energy to another.** A generator transforms mechanical energy into electrical energy. A steam or gas turbine transforms heat energy into mechanical energy.
2. **To transfer energy from one place to another.** See the example in Fig. 1-11. An automobile's connecting rods, crankshaft, drive shaft, and rear axle transfer energy. It is transferred from the combustion in the cylinders to the rear wheels.
3. **To change force and distance relationships.** A small force through the action of a machine can move a heavy object. Also, distance relationship can be changed. A person can cover more distance in a given period of time on a bicycle than by running.

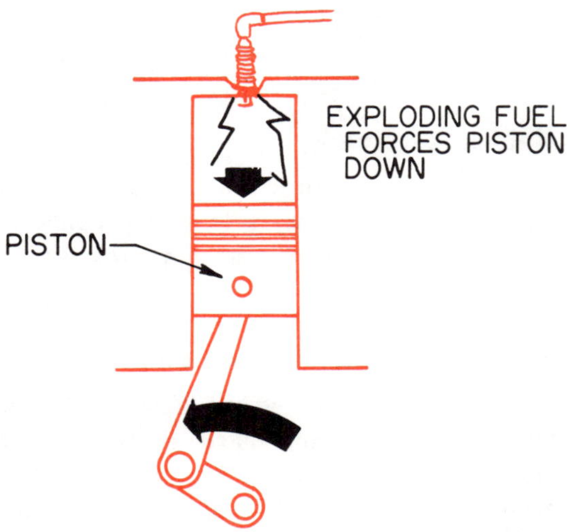

Fig. 1-11. Up and down motion transformed into rotary motion.

Many other sources of power come from nature. Three forms of energy received directly from nature are **water, wind,** and **sun.** Sources of energy which can be changed into power that produces work include wood, coal, oil, and gas. Each source of energy can be changed into different forms of force and power. For example, water can be used directly to do work. But it can also be heated to produce steam to provide added power.

People have used simple machines and energy sources to change the world. They have combined scientific principles with simple machines to produce complex machines. They have multiplied human muscle power thousands of times. Power has progressed from the simple water wheel to the steam engine. From there it has progressed to the internal combustion engine and then to the rocket engine. This is but one example of technology in action.

ENERGY

Energy is constantly present in three forms. They are **motion, heat,** and **light.** Besides these, energy is available in three other forms. They are electrical, chemical, and nuclear (atomic). Motion is given the more technical name of **mechanical energy.**

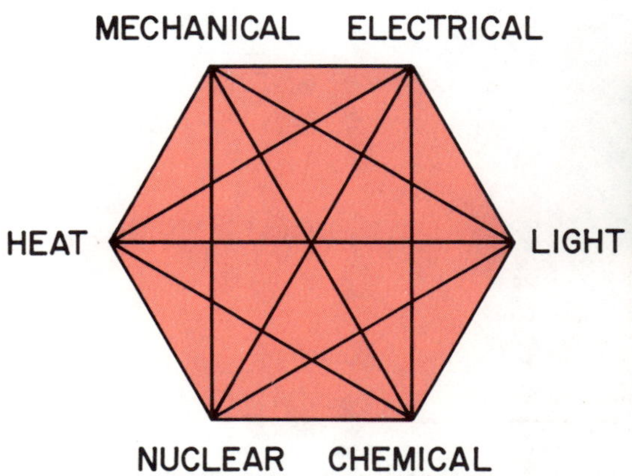

Fig. 1-12. All forms of energy are related.

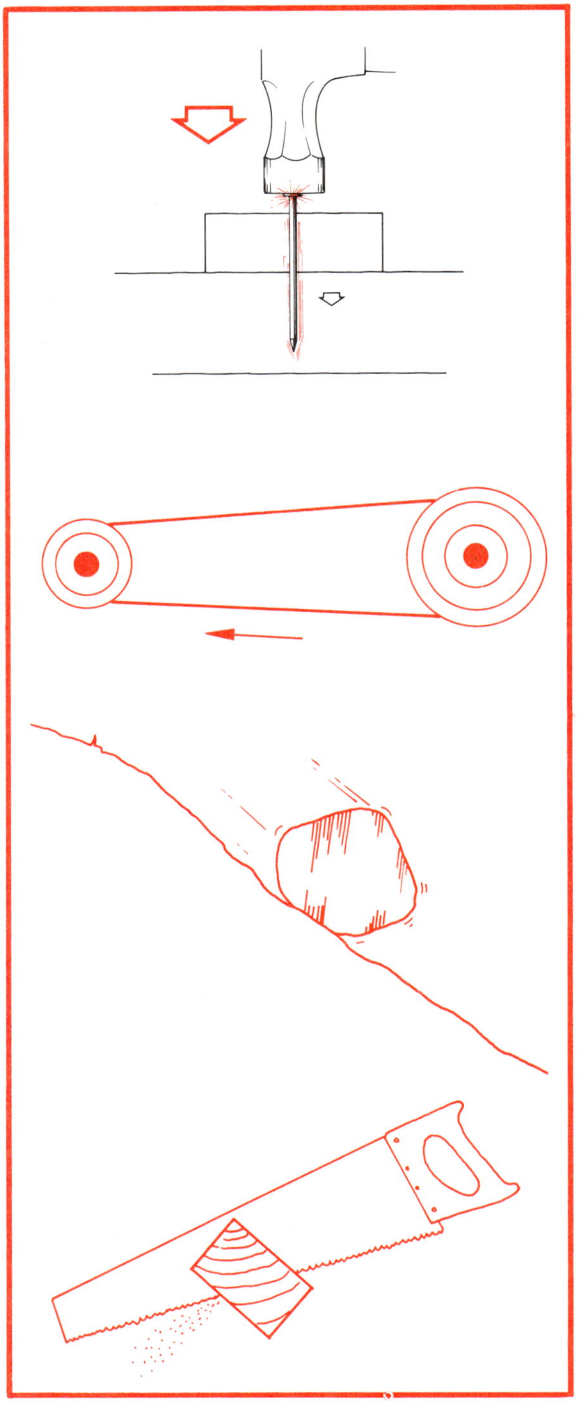

Fig. 1-13. Mechanical energy in the form of belt and pulleys, falling rock, and cutting of wood.

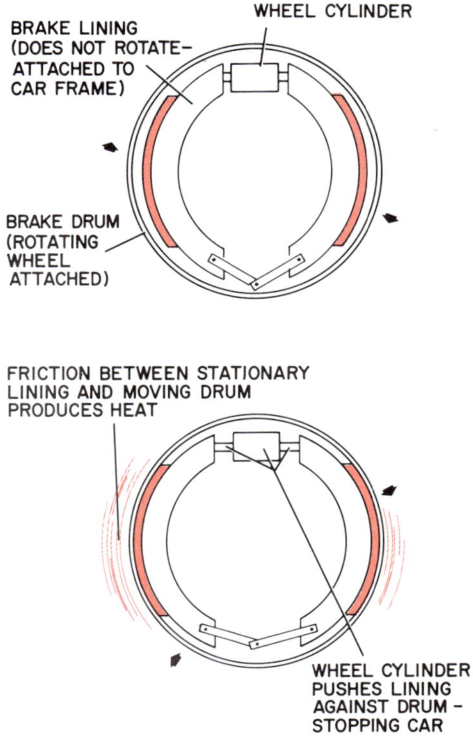

Fig. 1-14. Operation of automobile brakes.

These six forms of energy are all related, Fig. 1-12. One form can be transferred or changed into any other form. Some transfers are easy. A fire changes chemical energy into heat and light energy.

Mechanical Energy

Mechanical energy is the most familiar of the six forms of energy. It is **energy involved in motion.** Every moving object has mechanical energy, Fig. 1-13.

Friction and Energy Conversion

Mechanical energy can become destructive. But most mechanical energy can be disposed of in a harmless way. When you stop an automobile, you must overcome the mechanical energy (motion). To do this, you apply the brakes. The brakes consist of a rough material that is pressed against the rotating wheel. This action produces friction. See Fig. 1-14. **Mechanical friction is resistance to movement between two surfaces that touch each other.** Friction between the brake lining (rough surface) and brake drum (rotating wheel) stops the car. The rotating brake drum rubs against the rough material. A large amount of heat is developed. Air passing over the wheels removes the heat. As a result, mechanical energy is changed or transferred into heat energy. This heat energy goes into the air. The car is stopped.

Nuclear Energy

Nuclear energy is often thought to be an exception to the idea that energy cannot be created or destroyed. Nuclear energy is developed by **changing matter into energy,** Fig. 1-15. Matter is any material or substance, such as water, iron, and oxygen. The materials used to produce nuclear energy are hydrogen and uranium. Energy is developed by changing part of the center, or nucleus of each atom into energy. This change lowers the amount of matter. It produces energy equal to the amount of matter lost. A very small amount of matter changed in this way produces huge amounts of energy.

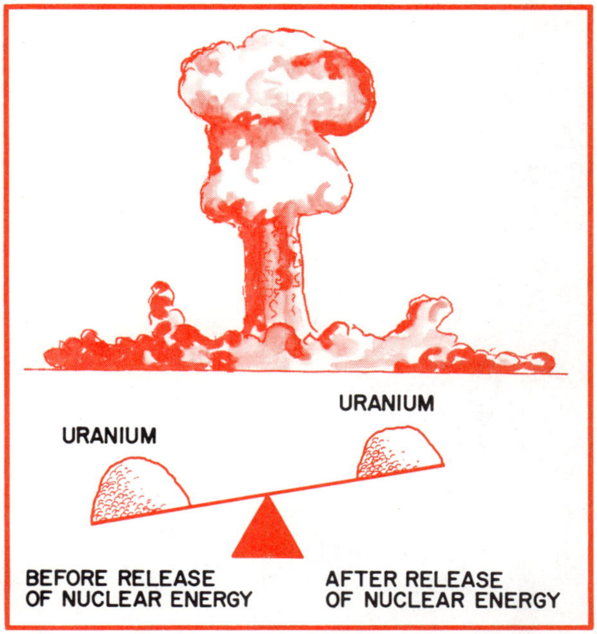

Fig. 1-15. Matter lost was changed into energy.

Heat Energy

Heat energy is a common form of energy, Fig. 1-16. Heat is used to develop mechanical energy. It can move an automobile, cook food, and generate steam to produce electricity. It can also heat homes and keep plants and animals alive.

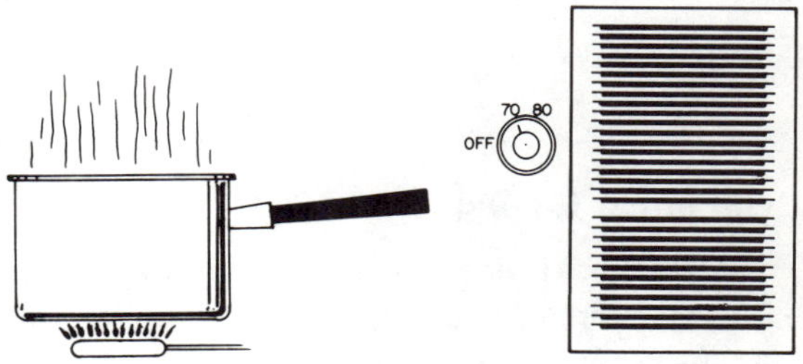

Fig. 1-16. Heat energy from gas flame and heater.

Light Energy

Light energy is able to do work. It has vast potential. Light energy floods the universe. It is produced by the sun and all other stars. Tiny amounts of light energy even come from stars that we cannot see.

Light energy promotes growth in plants and keeps alive the animals that eat them. It also serves people by making the surroundings on earth visible. Light energy from the sun and the rest of the universe can be called **radiant energy,** Fig. 1-17. Radiant energy brings both heat and light to the earth. This combined energy is often called the **energy of the universe.** It is the form of energy that reaches to the far corners of all of the known universe.

Radiant energy is very important to life on earth. It is the manner in which all energy is received from the sun. It is also the manner in which the earth loses excess energy.

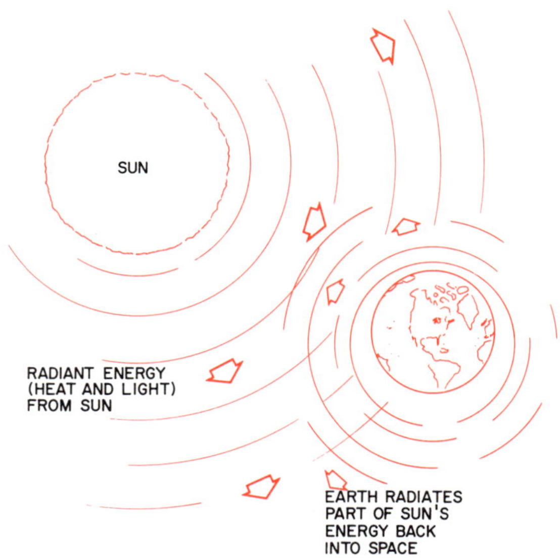

Fig. 1-17. Earth acts as a receiver and transmitter of radiant energy.

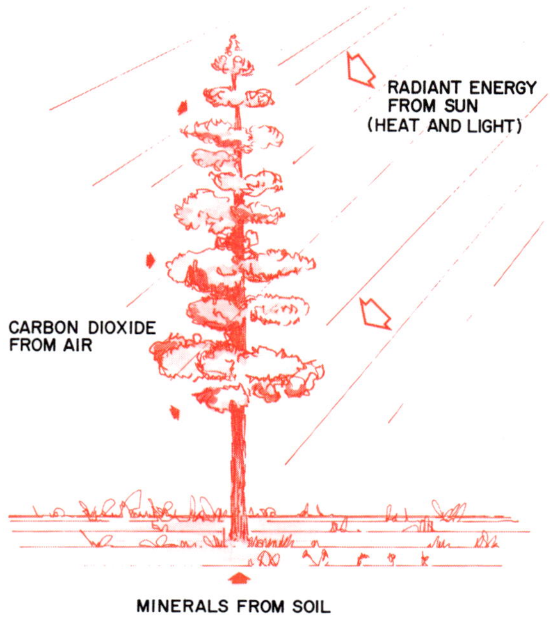

Fig. 1-18. Radiant energy is changed into chemical energy.

Chemical Energy

Chemical energy is the **energy of life.** Plants use radiant energy from the sun to grow. As the plants grow, the sun's energy is changed to chemical energy. The plants are fed by water from the earth and carbon dioxide from the air. Minerals from the soil aid the change to chemical energy. See Fig. 1-18.

When heat is needed, the chemical change is reversed by burning the plant. The energy stored by the plant is given back in the form of heat and light, Fig. 1-19.

There are various types of chemical energy. The common flashlight battery produces electricity. It does this by changing chemical energy into electrical energy. An explosion is also a release of chemical energy. In an explosion, the chemicals are changed into heat, light, and motion (wind and pressure).

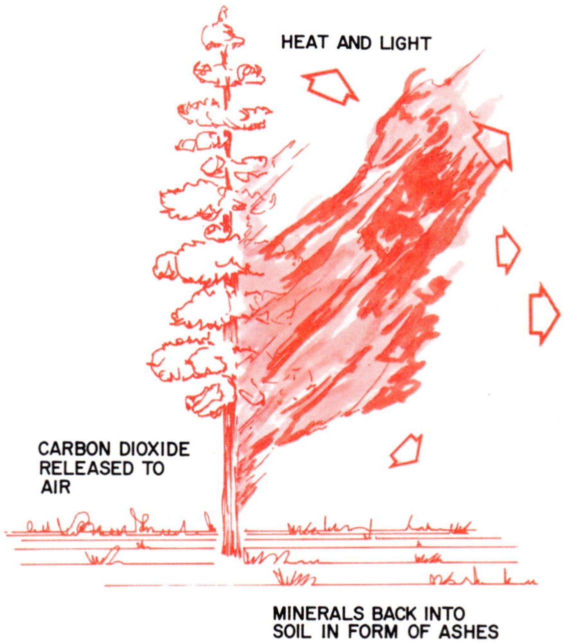

Fig. 1-19. Chemical energy is changed into heat and light (radiant energy) by burning.

Electrical Energy

Electrical energy is the most flexible form of energy. Electrical energy is the motion of tiny particles of matter called **electrons.** An electron is too small to see. So electricity must be studied by looking at the changes it produces.

POWER MECHANICS

A complete study of energy and its uses includes almost everything around you. It also includes the entire universe and its actions.

This book is about power mechanics. It is just one small but important part of the total study of energy. Power mechanics is the study of the generation and conversion (the producing and changing) of energy to power. It is also a study of the transmission, control and use of this power.

This study will focus on mechanical energy in the form of motion. You can learn about the ways in which this motion is put to useful work. Power mechanics includes the following:

1. Development of power for industrial and home use.
2. Control and measurement of power.
3. Transmission of power through mechanical, fluid, and electrical means.
4. Use of power to do work.

CAREER OPPORTUNITIES IN POWER MECHANICS

Power mechanics is not a career within itself. You will find that power mechanics is a group of many kinds of work. It is sometimes called a **family of occupations.** You may find a career in power sources, transportation, transmission and control, or use of power. Careers in these fields cover all levels of interest and ability.

Careers vary from production and assembly of power devices to research in future power sources. In between are hundreds of positions. They include engineers, industrial technologists, technicians, tradesmen, and service personnel. Also, there are many careers in related fields, such as sales, public relations, and carrying on business.

Careers in Generating Electricity

Power sources are fast changing. Nuclear power may soon provide 20 percent of all electrical power. Solar (sun) energy can be changed directly into electrical energy. It will soon provide more heat for homes and other buildings. This is a growing source of power and will provide many people with future careers.

There are many chances for careers in power mechanics. There are careers in research and engineering. Workers here may plan and design better nuclear reactors. They may direct the operation of large electrical plants. Some workers supervise or direct work at a plant. Others work in the transmission of electricity from source to use. Many other workers are needed to construct, maintain, and repair the huge systems of electrical plants and transmission lines. See Figs. 1-20 and 1-21.

General Electric Company

Fig. 1-20. A technician operates a modern plant from this control room.

U.S. Bureau of Reclamation

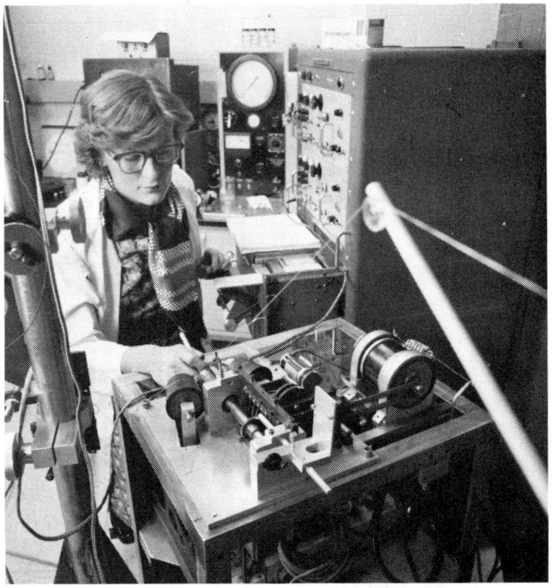

Fig. 1-21. An engineer checks the operation of new fuel refinery equipment.

Careers in Transportation

Transportation of people and goods is another of the giant families of careers found in power mechanics. See Figs. 1-22 to 1-25. You can find many careers in automotive, aircraft, railroads, and shipping. The problems of transporting people and goods are more difficult because of air pollution. Heat engines are one of the main sources of impure air. The problem must be solved.

The future of transportation may bring about many new career fields. This can happen when power sources are found to replace the heat engines now used.

Careers in Energy Sources

Energy resources is another family of careers. It includes the companies that now supply all of the fossil fuels in use. People in research are trying to find ways to replace these fuels. But fossil fuels will likely supply most of the power during the rest of this century. Most

Douglas Aircraft Company

Fig. 1-22. Workers operate overhead cranes to lower an airplane fuselage onto a wing section.

Ford Motor Company

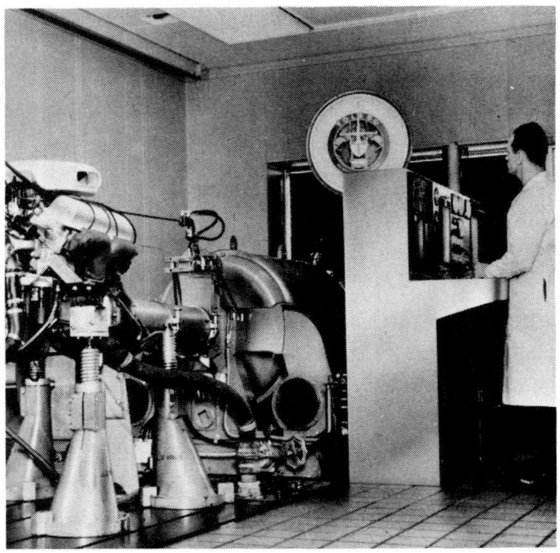

Fig. 1-23. A technician supervises the testing of a modern automobile engine.

careers are in the recovery, refining, and transportation of fossil fuels.

More jobs are opening in different aspects of solar energy, nuclear energy, and chemical energy. Careers may be found anywhere from maintenance and service to research engineer. See Figs. 1-26 and 1-27.

The Boeing Company

Fig. 1-24. Engineers check procedures for testing the turbofan jet engine suspended just behind them.

Shell Oil Company

Fig. 1-26. A research chemist studies the chemical and physical properties of a new fuel.

Piper Aircraft Corp.

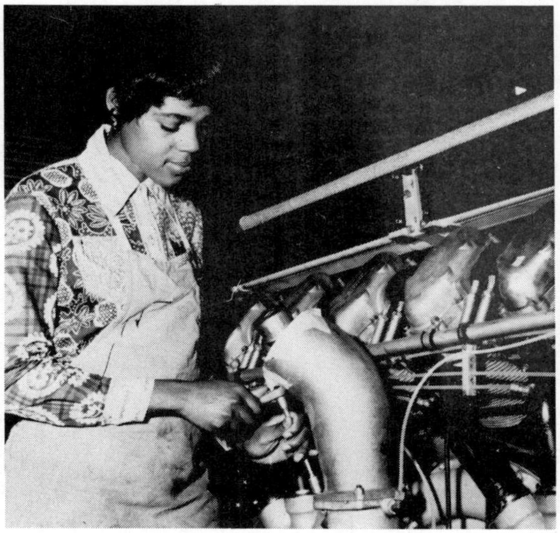

Fig. 1-25. A diesel mechanic adjusts diesel equipment for safe operation.

Shell Oil Company

Fig. 1-27. Engineers study the blueprints of a modern gasoline refinery.

Careers in Transmission and Control

Transmission and control is one of the fastest growing career fields in the nation. **Automation** has developed through the progress made in the control and transmission of power. Automation is the production of goods and products by self-controlled machines. Computers direct and control the operations. Instruments are needed to check and correct operating procedures. Instruments also provide information to computers. This information, called **feedback,** reports on the progress and accuracy of each operation. The motion and actual manufacturing is provided by three power devices. They are fluid, mechanical, and electrical. These are parts of an automated system. They may be thought of as the brains, senses, and muscles of automation.

Do not think a computer has mental ability. The computer, like all automation, is the product and servant of the scientist, engineer, and technician. The program planner and engineer plan what the computer does. People are needed for the operation of automated plants. As this field grows, there will be fewer jobs that require low skills. As a result, better educated workers are needed to fill the new technical jobs that develop.

Combustion Engineering

Fig. 1-28. Combustion engineers design and test equipment that supplies steam power to generate the world's electricity.

Pacific Gas and Electric Company

Fig. 1-29. A mechanical engineer plans new gas lines with the help of a pipe line map and a slide rule.

Chapter 2 MECHANICAL POWER

Mechanical power is the transmission and control of motion through the use of gears, pulleys, shafts, and other devices. A mechanical power system may be divided into four parts:

1. Input.
2. Control.
3. Transmission.
4. Output.

An automobile is a good example of a mechanical power system. It contains each of the four parts of the complete system. **Input** is the form in which motion is produced and placed in the mechanical system. In the automobile, input is the motion produced when gasses expand in the combustion chamber of the engine. See Fig. 2-1.

Output is the motion taken from the system. In the automobile, output is the revolving of the rear wheels.

Between input and output, the motion is changed through the use of various **control**

Ford Motor Company

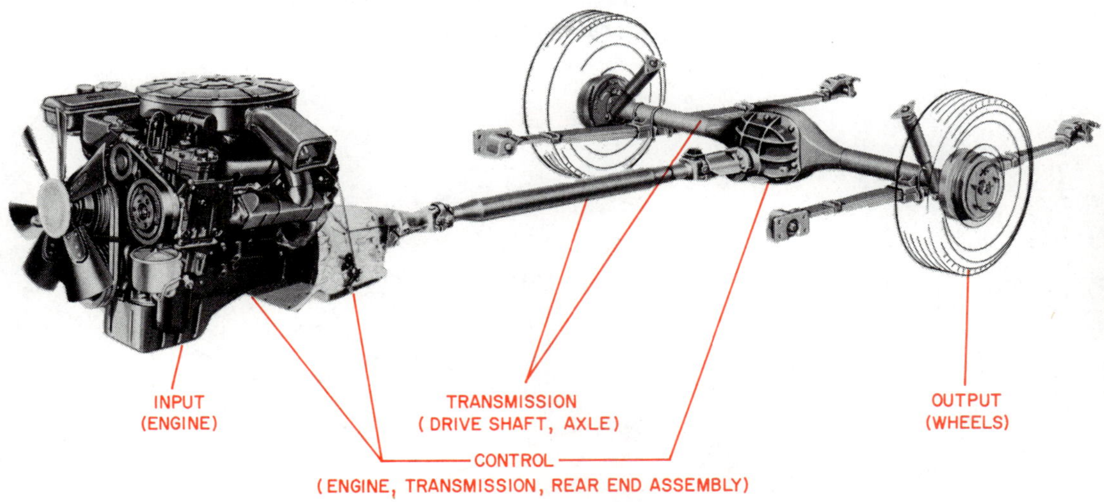

Fig. 2-1. The automotive drive line is a mechanical power system.

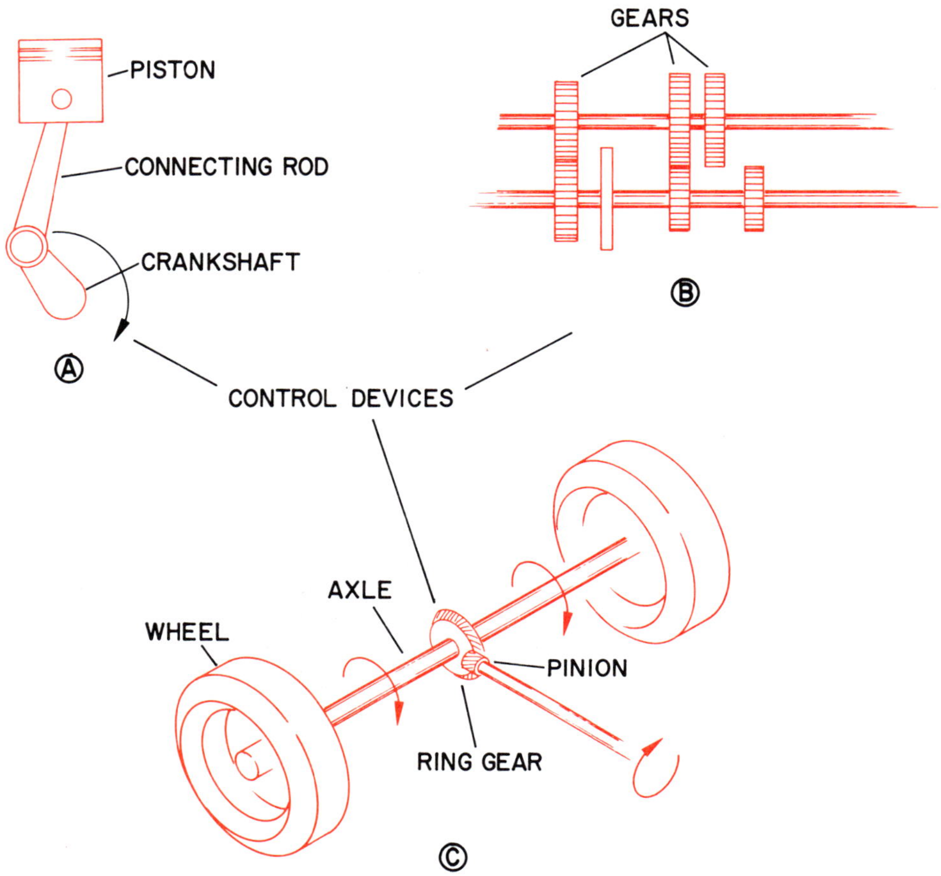

Fig. 2-2. Power control occurs throughout the system.

A. Engine changes downward push into rotary motion.

B. Transmission changes speed and direction of motion.

C. Rear end assembly changes speed and force and changes direction of motion.

Chrysler Corp.

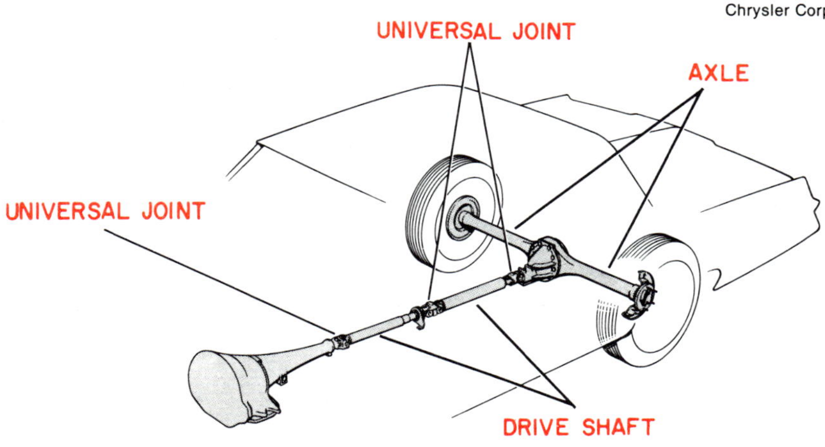

Fig. 2-3. Power is transmitted through rotating axle and drive shafts and universal joints.

devices. See Fig. 2-2. The piston has an up and down motion. This motion is changed to rotary motion by the connecting rod and crankshaft (A). Speed and direction of the tires' rotation may be changed by the automobile's transmission (B). The rear end assembly always reduces the speed. It also increases the torque (force) and turns the rotating motion 90° (C). Last, the change from crankshaft rotation to forward motion occurs by the action of the wheel on the ground. The size of the wheel determines the distance the car moves during each revolution or complete turn of the axle (C).

After the motion has been developed (input), it must be carried to where it will be used. This is the **transmission phase.** In the automobile, motion is transmitted by the drive shaft, universal joints, and rear axles. This is shown in Fig. 2-3. As power is transmitted, many control devices change the form of the motion. As a result, control is usually a part of the **transmission of power.**

POWER INPUT

Power for the mechanical system normally comes from either of these sources:

1. An engine operating on gasoline or some other fuel. The automobile engine, the airplane jet engine, and the small engine that powers a lawnmower all operate on gasoline or other fuels made from oil.
2. A motor operating on electricity or fluid power. The drill press motor and the air hammer motor are two of many types of motors that use electricity, air, or oil to produce motion.

Each of these sources produces motion. The motion produced is one of three types. See Fig. 2-4.

1. Reciprocating.
2. Rotary.
3. Linear.

Fig. 2-4. Input motion is in one of three forms: reciprocating, rotary, or linear.

Reciprocating, Rotary, and Linear Motion

Reciprocating motion is a back and forth motion. An example is the piston of the engine. This motion is produced by most gasoline and diesel engines.

Reciprocating motion has very few direct uses. It is usually changed to **rotary motion** within the engine. The power produced by a reciprocating engine is used to make rotary motion. See Fig. 2-5.

The jet engine is a major exception to the use of rotary motion. The jet engine produces a forward thrust or force. This force results in **linear** or **straight line** motion. It is opposite to the escaping exhaust gases. It is put to direct use in propelling a jet plane.

Control

Motion produced by an engine, turbine, or electric motor may be put to use without change or control. The rotary motion of the steam turbine can be used to rotate generators. The rotary motion of small gasoline engines can be used to drive the rotating blade of a rotary lawnmower.

However, most devices in industry and homes need to change motion. The change may be an increase in speed. It may be an increase in force. It may be a change in direction or a start and stop action. Whenever change is needed, controls that produce this change are used.

BASIC MACHINES

Before you study control devices, you need to understand the basic principles of their operation. You can begin by learning about the operation of basic machines. They are often called the **six simple machines.** These

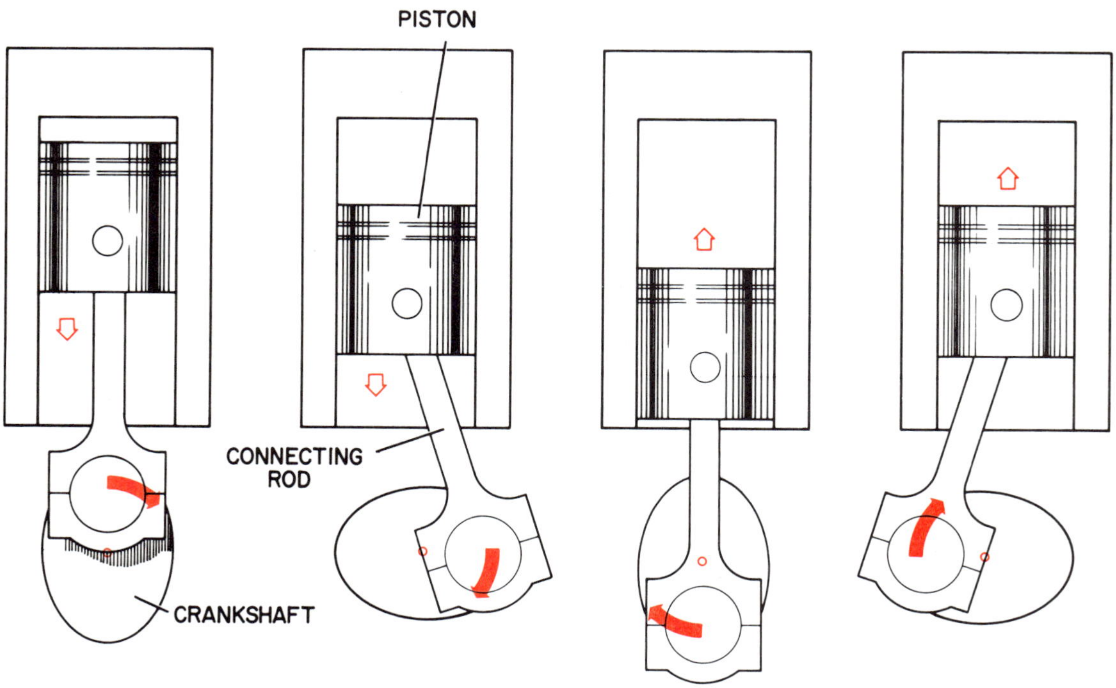

Fig. 2-5. Reciprocating motion is changed to rotary motion by the combined action of the connecting rod and crankshaft.

machines are the lever, wheel and axle, pulley, inclined plane, wedge, and screw. Their major purpose is to change the force and speed produced by a person or machine.

Mechanical Advantage

Simple or basic machines provide a mechanical advantage. They allow a worker to lift a 500 pound engine from a car and place it on a workbench. See Fig. 2-6. In this example, a chain hoist provides the mechanical advantage. With the hoist, a worker may exert a force of 20 pounds through a distance of 50 inches. This might raise the engine only 2 inches. The mechanical advantage is 25 to 1. For every pound of force applied, the worker lifts 25 pounds of engine. By applying 20 pounds of force, a 500 pound engine can be moved.[1] (Note the metric conversion of this problem below.)

Using a simple machine gives you a gain in mechanical advantage, but there is **no gain in work.** Input work and output work are the same.

Input Work		**Output Work**
Force × Distance	=	Force × Distance
20 lbs. × 50 in.	=	500 lbs. × 2 in.
1000 inch-pounds	=	1000 inch-pounds[2]

Mechanical advantage is gained at a price. In this case, the price is distance moved. The 20 pounds must be applied over a distance of 25 inches to raise the engine 1 inch.[3] As you study the mechanical advantage of simple machines, watch for a loss in distance moved.

The fact that a change in force always brings about a change in distance can be put to use. The force-distance relationship can be used to change distance and speed. The rear wheel of a bicycle rotates faster than the pedals. This is done by using a chain and different size sprockets. The price of increased speed is an increase in effort needed to push the pedals. Work input and work output will be the same, just as with the chain hoist.

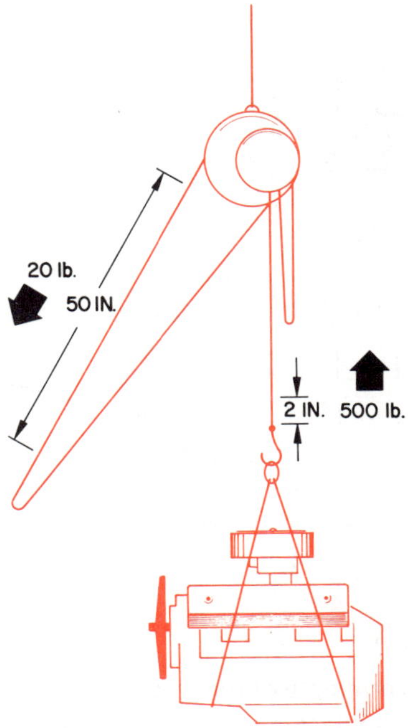

U.S. Customary	SI Metric
20 lbs.	9 kg
500 lbs.	225 kg
2 in.	50 mm
50 in.	1250 mm

Fig. 2-6. Mechanical advantage is produced by a chain hoist.

SI Metric Conversion

1. A worker may exert a force of 9 kilograms through a distance of 1250 millimeters. This may raise the engine only 50 millimeters. The mechanical advantage is 25 to 1. For every kilogram of force applied, the worker lifts 25 kilograms of engine. When 9 kilograms of force is applied, a 225 kilogram engine can be moved.
2. $\dfrac{\text{Input Work}}{\text{113 joules}} = \dfrac{\text{Output Work}}{\text{113 joules}}$
3. The 9 kilograms must be applied over a distance of 1250 millimeters to raise the engine 50 millimeters.

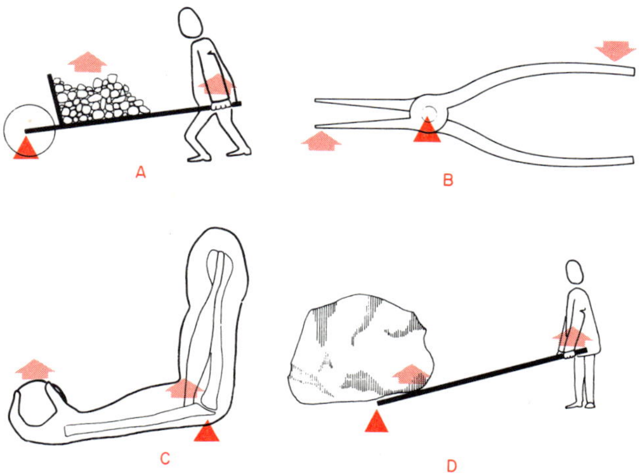

Fig. 2-7. Uses of the Lever

The Lever

The principle of the lever is used in **three of the six simple machines.** The wheel and axle and the pulley are changes in the principle of the lever.

Figure 2-7 shows different uses of the lever. In operation, the lever is very simple. It often serves two purposes. First, it provides mechanical advantage. Second, it changes the direction of the applied force. Figure 2-8 shows the three kinds of levers.

The mechanical advantage of the **lever** is easy to determine. Refer to Fig. 2-9. The mechanical advantage is a ratio between the lengths of the lever arms. In this example, the ratio is 5 to 1. As a result, only a 2-pound force is needed to balance the 10-pound weight.[4] Note that the product of force times distance is the same for both sides of the lever.

Wheel and Axle

The wheel and axle can be thought of as a **round lever.** See Fig. 2-10. It provides

SI Metric Conversions

4. As a result, only 9 newton force is needed to balance the 45 newton force.

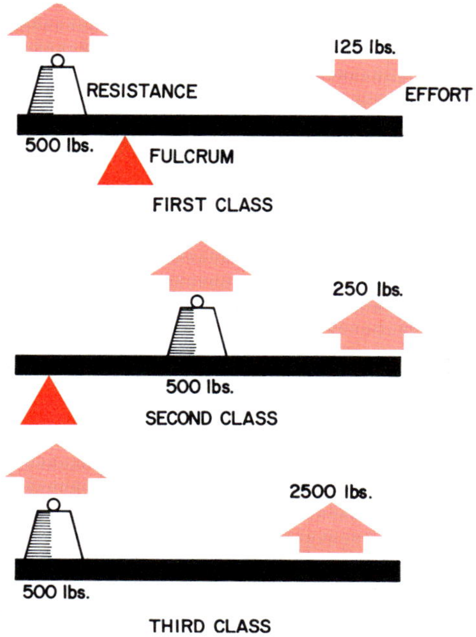

U.S. Customary	SI Metric
125 lbs.	57 kg
500 lbs.	227 kg
250 lbs.	113 kg
2500 lbs.	1134 kg

Fig. 2-8. Three Classes of Levers

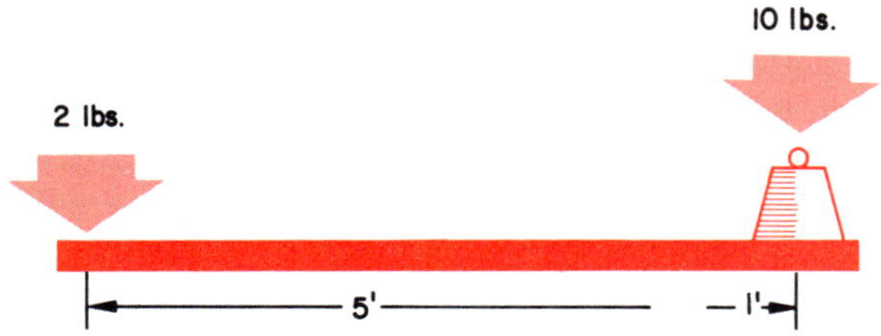

U.S. Customary	SI Metric
2 lbs.	9 N
10 lbs.	45 N
1 ft.	0.3 m
5 ft.	1.5 m

Fig. 2-9. This first-class lever provides a mechanical advantage of 5 to 1.

Input = Output
Force × Distance = Force × Distance
2 lbs. × 5 ft. = 10 lbs. × 1 ft.
10 Pounds-Feet = 10 Pounds-Feet

SI Metric Conversion

9 N × 1.5 m = 45 N × 0.3m
13.5 newton-meters = 13.5 newton-meters

mechanical advantage in the same way as a second class lever. This is done by applying the force to the large wheel.

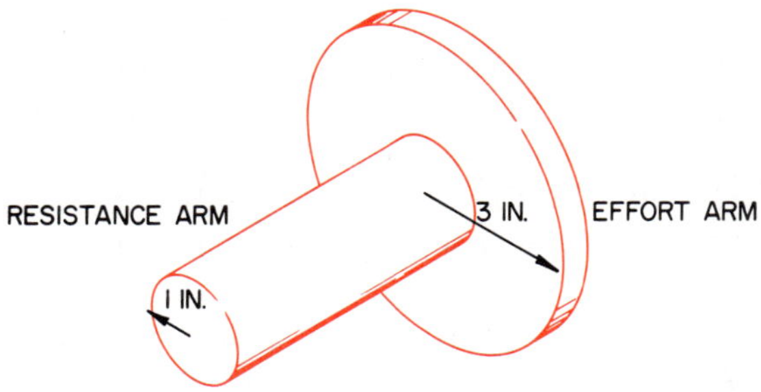

U.S. Customary	SI Metric
1 in.	25 mm
3 in.	76 mm

Fig. 2-10. The mechanical advantage of the wheel and axle is the ratio between the two lever arms, which is 3 to 1.

Pulleys

Pulleys are used to change the direction of an applied force. They also provide mechanical advantage. Pulleys are often used to lift heavy weights.

Figure 2-11 shows the way in which pulleys gain mechanical advantage. Pulleys have the same relationship between force and distance moved as do levers and the wheel and axle. In **A** of Fig. 2-11, a downward pull on the rope produces an equal upward movement of the weight. This simple pulley changes the direction of motion, but it gives no mechanical advantage. Mechanical advantage is written as a ratio. When there is no mechanical advantage, there is a ratio of 1 to 1 (1:1).

In **B** of Fig. 2-11, the pulley now moves with the weight. There is no directional change. Note that the weight is supported by two ropes. As a result, the weight is equally divided between

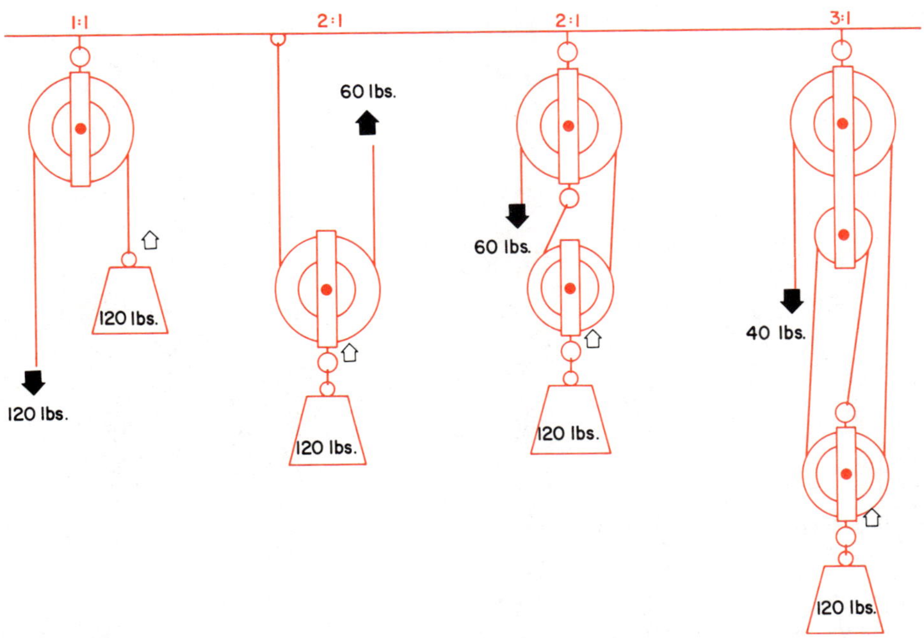

Fig. 2-11. Mechanical advantage can be gained through the use of pulleys.

U.S. Customary	SI Metric
40 lbs.	178 N
60 lbs.	267 N
120 lbs.	534 N

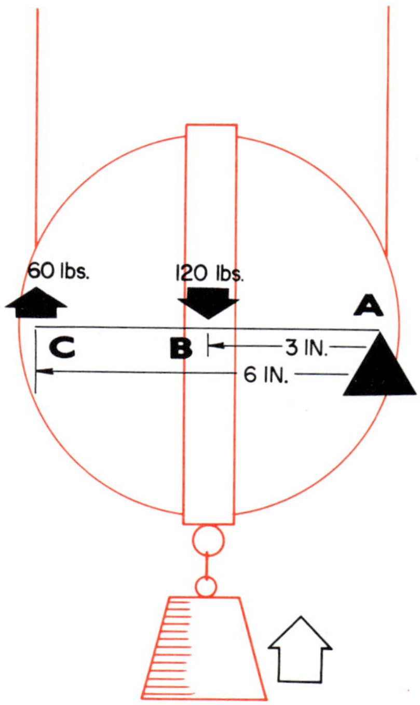

U.S. Customary	SI Metric
60 lbs.	267 N
120 lbs.	534 N
3 in.	75 mm
6 in.	150 mm

Fig. 2-12. The pulley operates as a second-class lever.

the two supporting ropes. Each holds 60 pounds of the 120-pound weight. When a force of 60 pounds is applied to the rope, the weight will move upward.[5] The rope will move twice the distance that the weight moves. This pulley system provides a mechanical advantage of 2 to 1.

In view **C,** a fixed pulley is added to the pulley shown in view **B.** This provides a change of direction. But it does not change the 2 to 1 mechanical advantage.

In view **D,** a third pulley is added. Since three ropes now support the 120 pound weight, each rope supports only 40 pounds. The mechanical advantage is 3 to 1. A force of 40 pounds must be applied a distance of 3 feet to move the weight up 1 foot.[6]

The pulley uses the principle of the second class lever. This is shown in Fig. 2-12. The fulcrum or pivot point is at point **A.** The weight from the center of the pulley is at **B.** Force is exerted at point **C.** The distance from the fulcrum

SI Metric Conversions

5. Each rope supports one half of the weight. When a force of 267 N is applied to the rope, the weight moves upward.
6. Each rope supports one third of the weight. The mechanical advantage is 3 to 1. A force of 178 N must be applied a distance 0.9 m to move the weight up 0.3 m.

Fig. 2-13. The mechanical advantage of 5:1 is produced by an inclined plane.

U.S. Customary	SI Metric
50 lbs.	220 N
250 lbs.	1100 N
2 ft.	0.6 m
10 ft.	3 m

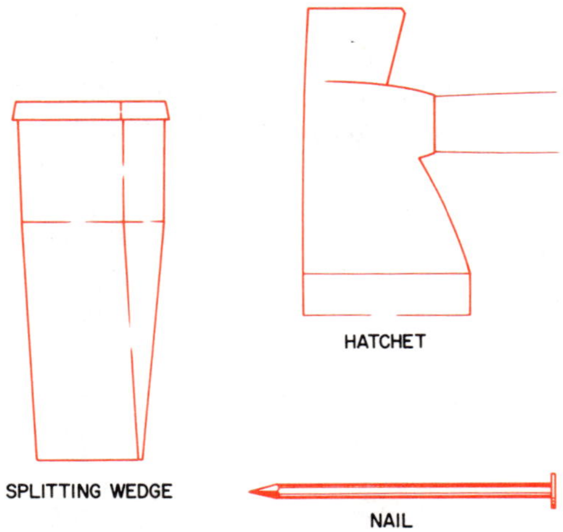

Fig. 2-14. Uses of the wedge.

to the applied force is always twice the distance to the load. As a result, a mechanical advantage of 2 to 1 is produced.

Inclined Plane

The inclined plane and the lever are the most important machines used in industry. The other four simple machines use the principles of these two. The **wedge** and the **screw** are adapted from the **inclined plane.**

The mechanical advantage of the inclined plane is shown in Figure 2-3. As the weight is pushed up the incline, a constant force must be exerted. This force is determined by comparing the distance moved **forward** to the distance moved **up.** It is difficult to lift a 250 pound barrel and put it on a platform. See Fig. 2-13. But it is easy to get the barrel on the platform by using the inclined plane. With the mechanical advantage of the inclined plane, the barrel can be rolled 10 feet with a force of 50 pounds.[7]

Wedge

The **wedge** is so much a part of the inclined plane that it is often not thought of as a separate machine. The wedge is just a small inclined plane used as a tool. Mechanical advantage is figured in the same way as for the inclined plane.

The wedge uses mechanical advantage to multiply a small force. The increased force is put to work. Examples of this are shown in Fig. 2-14. The splitting wedge multiplies the blow of a hammer to split a log. The nail uses force developed by the wedge to pierce the board.

SI Metric Conversion

7. With the mechanical advantage of the inclined plane, the barrel can be rolled 3 m with a force of 220 N.

Screw

The **screw** is simply an inclined plane wrapped around a shaft. See Fig. 2-15. A very gentle slope is given to the threads of a screw. This makes a very high mechanical advantage possible. A ½ inch bolt with 20 threads per inch has a mechanical advantage of more than 30 to 1. When the threads are unwound, as shown in Fig. 2-15, they will make an inclined plane over 30 inches long for each inch of movement on the bolt.[8] Figure 2-16 shows different uses of the screw.

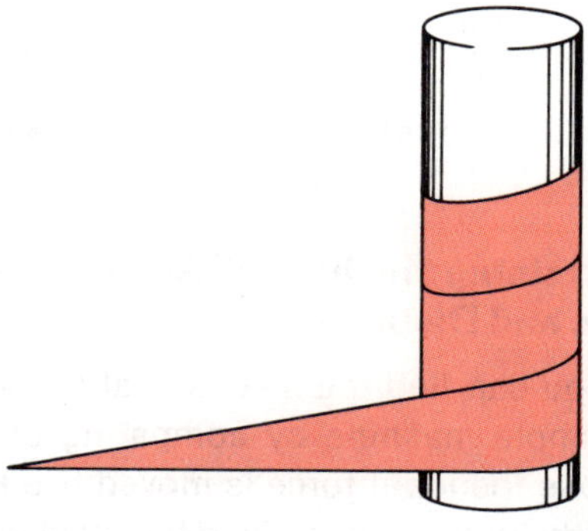

Fig. 2-15. The principle of the screw is shown by an inclined plane wrapped around a shaft.

SI Metric Conversion

8. A 12 mm bolt with 1 thread per millimeter has a mechanical advantage of over 30 to 1. When threads are unwound, they will make an inclined plane over 30 mm long for each millimeter of movement on the bolt.

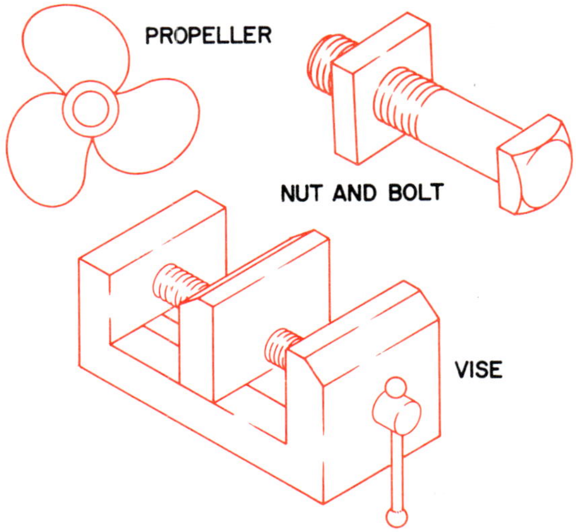

Fig. 2-16. Different uses of the screw.

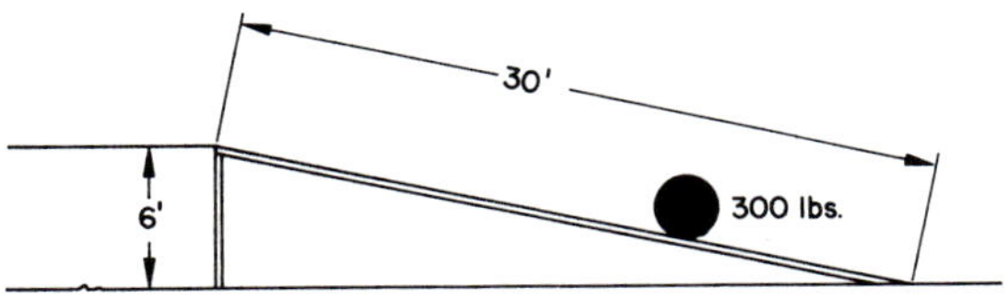

U.S. Customary	SI Metric
300 lbs.	1335 N
6 ft.	1.8 m
30 ft.	9 m

Fig. 2-17. Moving a barrel up an inclined plane.

Determining Mechanical Advantage, Force, and Distance

You can find the mechanical advantage of any simple machine by comparing distances. Suppose the input force is moved five times as far as the output force. Then the mechanical advantage is 5 to 1 (5:1).

SI Metric Conversion

9. A barrel is rolled a distance of 9 m on an inclined plane to raise the barrel 1.8 m.
10. The mechanical advantage is:

$$\frac{9\text{ m}}{1.8\text{ m}} = \frac{5}{1} \text{ or } 5:1$$

An example will show the calculations. A barrel is rolled a distance of 30 feet on an inclined plane to raise the barrel 6 feet.[9] What is the mechanical advantage? See Fig. 2-17.

$$\text{Mechanical Advantage} = \frac{\text{Input Distance}}{\text{Output Distance}}$$

$$\text{Mechanical Advantage} = \frac{30\ \cancel{\text{ft.}}}{6\ \cancel{\text{ft.}}}$$ [10]

$$\text{Mechanical Advantage} = \frac{5}{1} \text{ or } 5:1$$

CONTROL DEVICES

A mechanical power system has control devices to change the motion to a more usable form. These are most common changes:

1. On-off switching.
2. Direction.
3. Speed and force.

On-off switching is needed when the power output is stopped without stopping the input. The automobile engine is not turned off for red lights or stop streets. A clutch is used to disconnect the input (engine) from the power output (rear wheels). On-off switching is also used in industry, such as on presses and production lines.

Directional changes may be one of three types:

1. Power output may be reversed. The motion is then in the opposite direction. Backing an automobile is an example of reversing the power output.
2. Power may be turned. This might be the 90° turn produced at the rear end of the automobile, Fig. 2-2.
3. The last type of directional change is from linear to rotary (or the reverse). See Fig. 2-5. Changes of this type are often needed in home and industrial use. Most washing machines use a back and forth rotational movement of an agitator.

Speed and force are treated together. This is because the change of one produces an opposite change in the other. There is only one way that either speed or force can be changed without changing the other. That is by changing the amount of power being put into the system. Suppose an automobile travels at 50 miles (80 kilometers) per hour on a straight, level road. When the car is driven uphill, the driver steps on the gas pedal. Giving the engine more fuel increases the input power. The automobile now has the added force needed to climb the hill without the speed being reduced below 50 mph (80 kph).

Ohio Gear Company

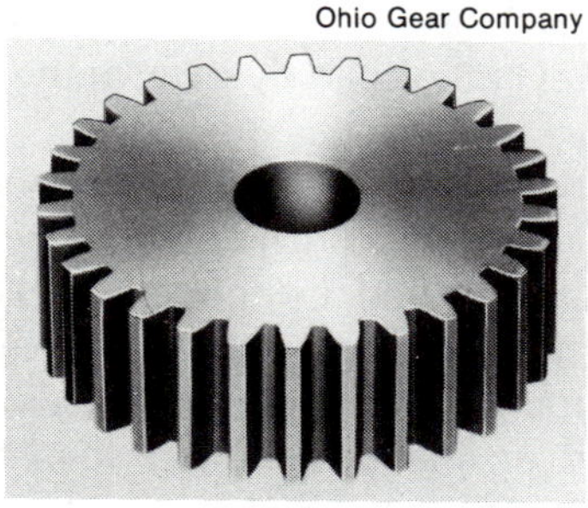

Fig. 2-18. In the spur gear, the teeth are parallel to the shaft opening.

Gears, pulleys and belts, clutches, and couplings are the most common devices used to produce these three changes in motion:

1. On-off switching.
2. Direction
3. Speed and force.

Gears

Gears may be used to provide directional change. They can be used to change speed and force. They can also provide for on-off switching. Many types and styles of gears meet the

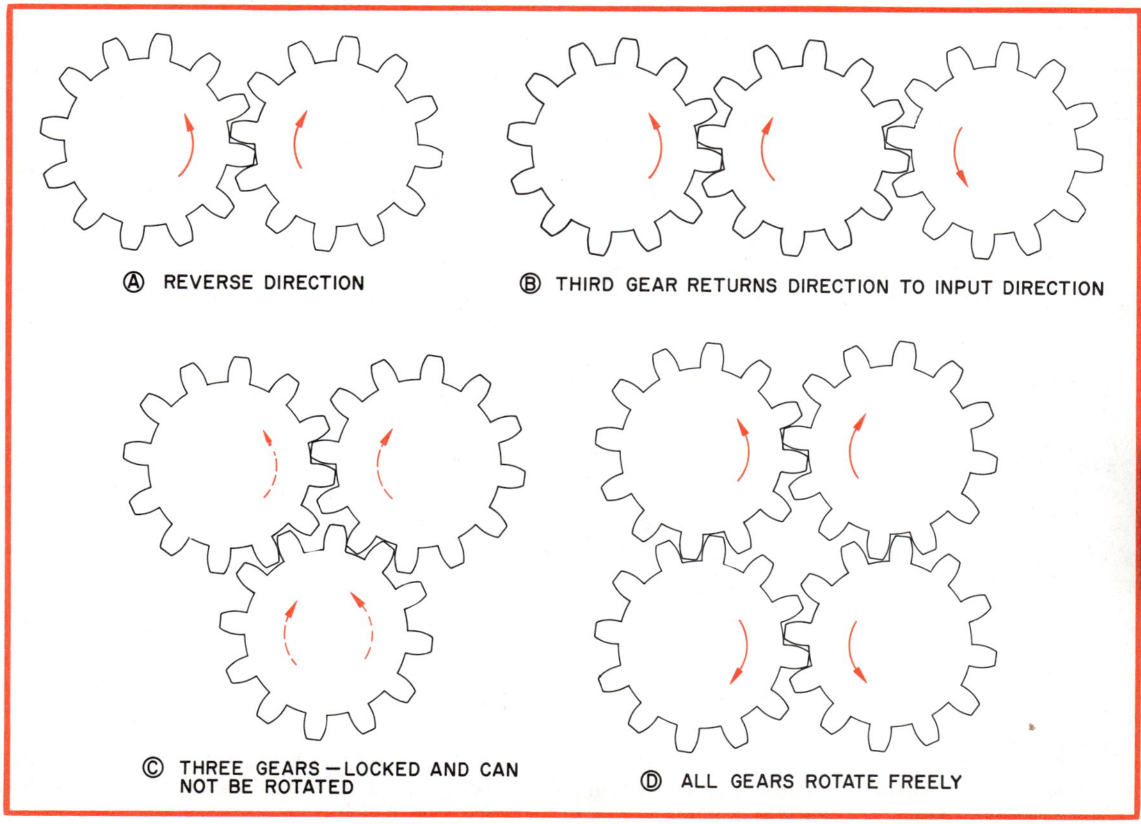

Fig. 2-19. Combinations of Spur Gears

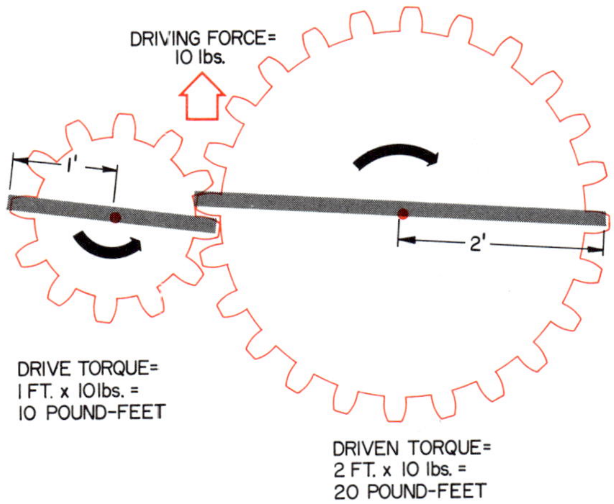

U.S. Customary	SI Metric
10 lbs.	45 N
1 ft.	0.3 m
2 ft.	0.6 m

SI Metric

Drive Torque = 0.3 m × 45 N = 13.5 newton-meters	Drive Torque = 0.6 m × 45 N = 27 newton-meters

Fig. 2-20. Torque is multiplied through the use of gears.

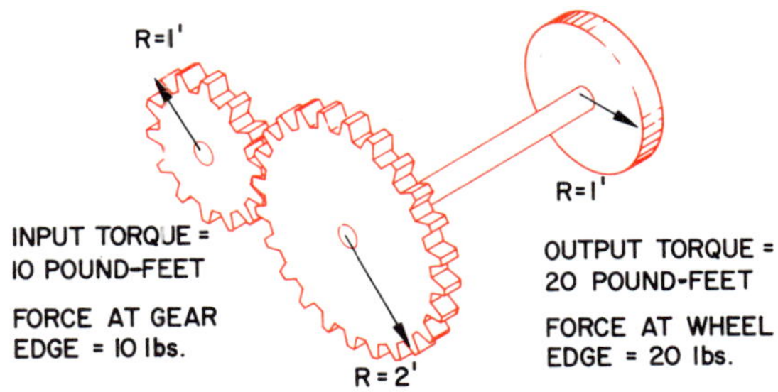

U.S. Customary	SI Metric
1 ft.	0.3 m

SI Metric

Input torque = 13.5 newton-meters	Output torque = 27 newton-meters
Force at gear edge = 45 N	Force at wheel edge = 90 N

Fig. 2-21. When a wheel and axle are added to the gears, the output force is multiplied.

special needs of industry. Three types of gears are discussed here:

1. Spur gears.
2. Helical gears.
3. Worm gears.

Spur Gear

The spur gear is the simplest gear design. See Fig. 2-18. The spur gear has teeth that are cut parallel to the shaft opening at the center of the gear.

Spur gears can provide directional change, Fig. 2-19. They also provide mechanical advantage. Spur gears use the principle of the lever to increase **torque.** Torque measures turning effort in pounds-feet. (In SI metrics torque is measured in newton-meters). It is found by multiplying the force by the distance from the center (radius). In this case, two levers are arranged as shown in Fig. 2-20. The end of one lever pushes on the end of the other lever. Gear teeth are substitutes for the lever ends. They extend all around each gear. The input gear is called the **drive gear.** The drive gear has a torque of 10 pound-feet. Since the gear has a radius of 1 foot, it is driving the second gear with a force of 10 pounds. This force is transferred to the driven gear. This means that the driven gear now has a force of 10 pounds pushing it in a circle. But the driven gear has a radius of 2 feet. Hence the torque is 10 pounds × 2 feet, or 20 pound-feet.[11] Note that the driven gear has twice as many teeth as the drive gear. As a result, the drive gear rotates twice for each revolution (or complete turn) of the driven gear. Torque was multiplied by 2, and the speed (number of revolutions) was divided by 2. A pro-

SI Metric Conversions

11. The drive gear has a torque of 13.5 newton-meters (Nm). Since the gear has a radius of 0.3 m, it is driving the second gear with a force of 45 N. This force is transferred to the driven gear. This means that the driven gear now has a force of 45 N pushing it in a circle. But the driven gear has a radius of 0.6 m. Hence the torque is 45 N × 0.6 m, or 27 newton-meters.

Morse Chain Company

Fig. 2-22. Helical Gears

Morse Chain Company

Fig. 2-23. Worm Gear Assembly
Each time the worm gear rotates once, the helical gear moves one tooth — 1/30 of a revolution.

portionate loss in speed went along with the gain in torque.

This torque can now be put to work. Figure 2-21 shows how the wheel can be used to put the torque to work at a greater force. (Note that the force at the outer edge of the gears of Fig. 2-20 remained constant.) This figure simply attaches a wheel and axle to the driven gear of Fig. 2-20. The 20 pound-feet of torque is transferred through the axle to the wheel. Since the wheel has only a 1 foot radius, the 20 pound-feet of torque produces a force of 20 pounds at the wheel's surface.[12] Since the wheel and axle

SI Metric Conversion

12. The 27 newton-meters of torque is transferred through the axle to the wheel. Since the wheel has only a radius of 0.3 m, the 27 newton-meters of torque produces a force of 90 N at the wheel's surface.

is only a form of the lever, multiplying this force is simply applying the principle of the lever.

Helical Gears. The helical gear is like the spur gear, but the teeth are cut at an angle. See Fig 2-22. The angle of the teeth gives a greater contact area. Usually, more than one tooth of a gear is engaged. With more contact area, more force is transferred. This helps keep the gears from wearing and breaking.

Worm Gears. Worm gears can make major changes in mechanical advantage and speed. See Fig. 2-23. Input into a worm gear unit is usually through the worm gear. The worm acts as a screw. It drives the helical gear as it rotates. A high mechanical advantage is possible. This is because the helical gear advances only one tooth each time the worm gear completes a full turn. Suppose the helical gear has 30 teeth. The worm gear will then rotate 30 times to rotate the helical gear once. This is a mechanical advantage of 30:1. It produces a torque 30 times greater than input torque.

Pulleys and Belts

Many kinds of devices are classified as pulleys and belts. Figure 2-24 shows some typical units. One group includes the chain and timing belt. It has **positive drive units.** The sprockets are toothed pulleys. Their teeth mesh with the teeth in the belt. The two drive sprockets are kept in the same relationship since the belt cannot slip. This is important in many uses in industry.

For example, in some gasoline engines a chain (or timing belt) and sprocket are used to drive the camshaft. This is shown in Fig. 2-25. The small sprocket is driven by the crankshaft. This sprocket, through the timing belt, drives the larger camshaft sprocket. The camshaft opens and closes the engine valves. These valves control the flow of fuel to the engine and must operate with the piston. When the piston is going down on the intake stroke, the intake valve must be open. Each valve action must be timed to the action of the piston. This means

B.F. Goodrich Company

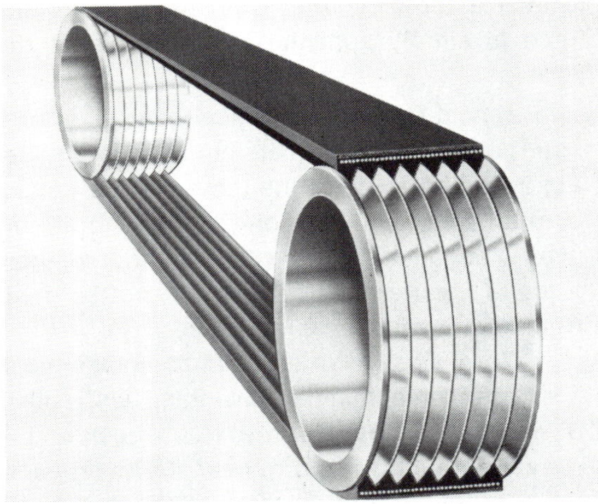

A. Multiple V-belt assembly, cutaway view.

Morse Chain Company

B. Roller chain assembly.

Morse Chain Company

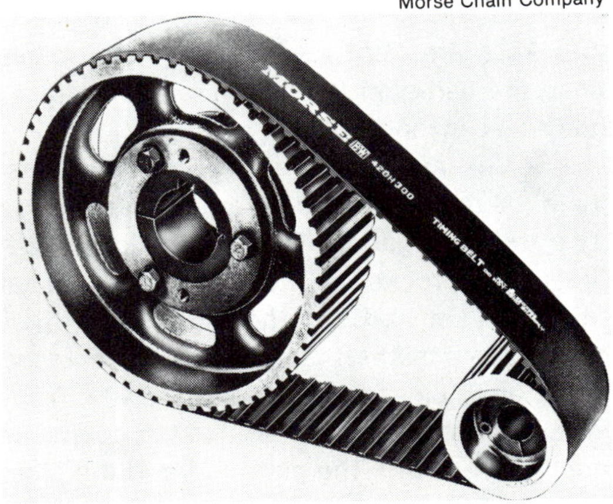

C. Timing belt assembly.

Fig. 2-24. Control and transmission of power through belts and chains.

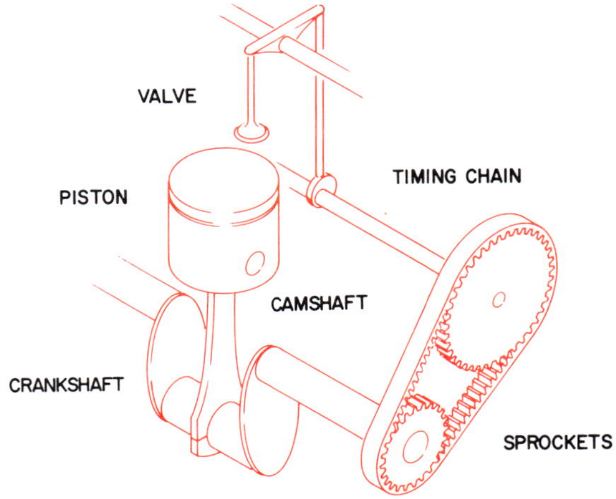

Fig. 2-25. The piston and valves must be timed so that the valves open and close at the correct time.

that the crankshaft and the camshaft must always remain in the correct timing with each other.

The other group has **non-positive drive.** In this group, the timing between the two units is not important. V-belts are used on the front of an automobile engine. These belts drive the engine water pump and alternator. The timing of these devices with the crankshaft is not important. As a result, simple V-belts and pulleys are used.

Pulleys and belts are able to provide for all three major control functions:

1. On-off switching
2. Directional change.
3. Speed and force change.

The first two are possible, but not often used.

On-off switching is possible by the loosening of one pulley. In this way, no power is transferred through the system. See Fig. 2-26. Directional change is possible by the twisting of the belt or use of idler pulleys. Neither of these systems is widely used.

The main use of the pulley and belt is to transmit power from one place to another. This often occurs with some change in force and speed. This system may be seen as two gears moved apart and connected by a belt or chain.

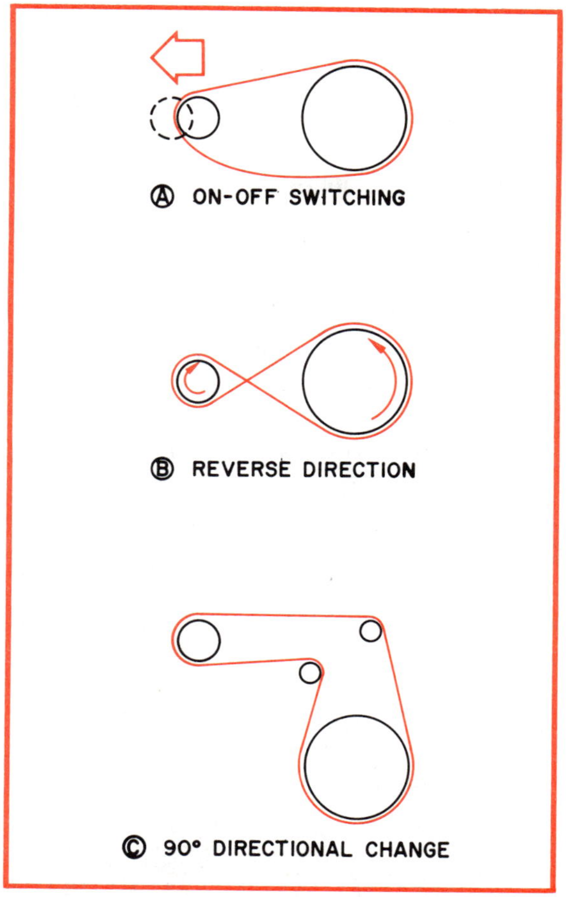

Fig. 2-26. Belts and pulleys can provide on-off switching and directional change.

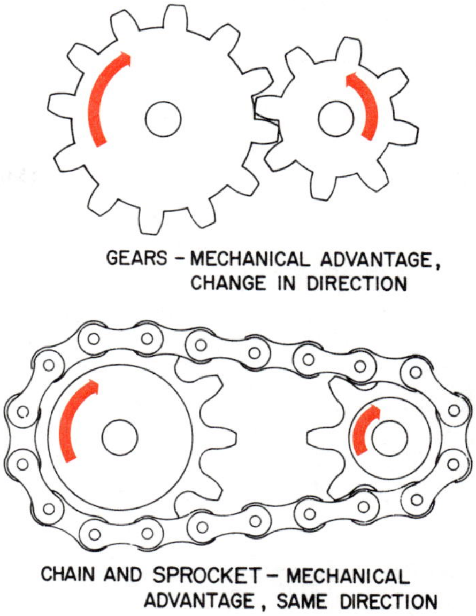

Fig. 2-27. Mechanical advantage of pulleys and belts, and chains and sprockets can be determined in the same manner as gears.

See Fig. 2-27. The ratios between the two pulleys provide mechanical advantage and speed changes in the same manner as gears. However, pulleys do not usually provide a change in direction or rotation as can be produced with gears.

Clutches

Clutches are designed to provide on-off operation. There are a large number of special purpose clutches. Only four of them are considered here.

1. Friction clutch.
2. Positive clutch.
3. Centrifugal clutch.
4. Overrunning or freewheeling clutch.

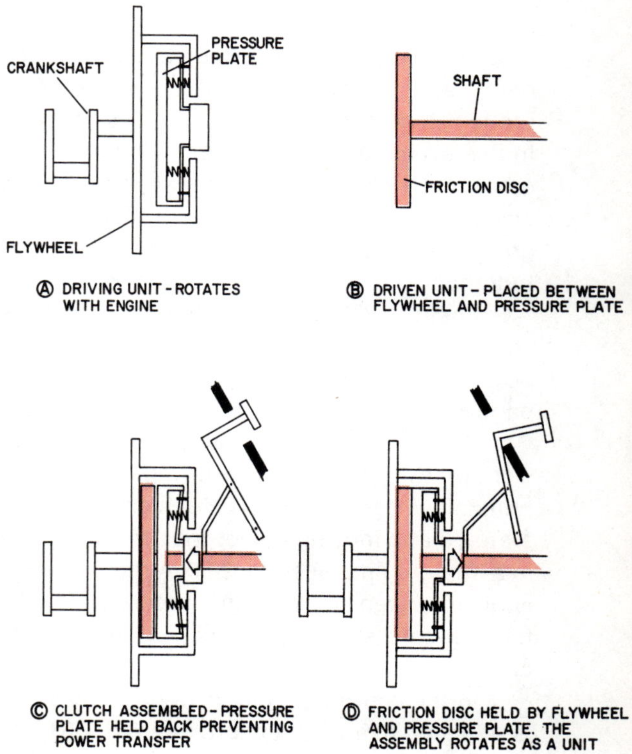

Fig. 2-28. The friction clutch used on many automobiles.

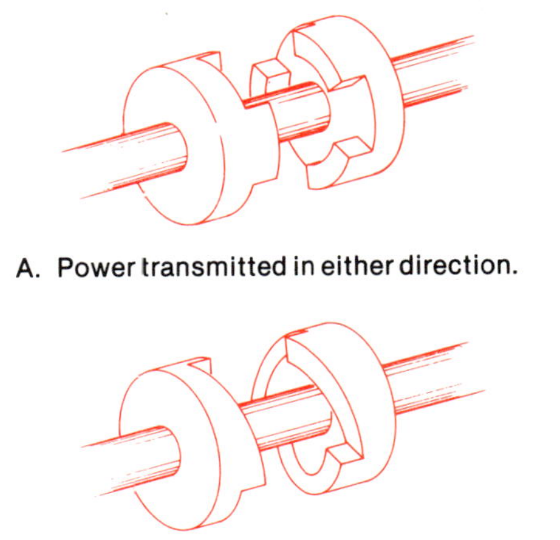

Fig. 2-29. Positive Drive Clutch

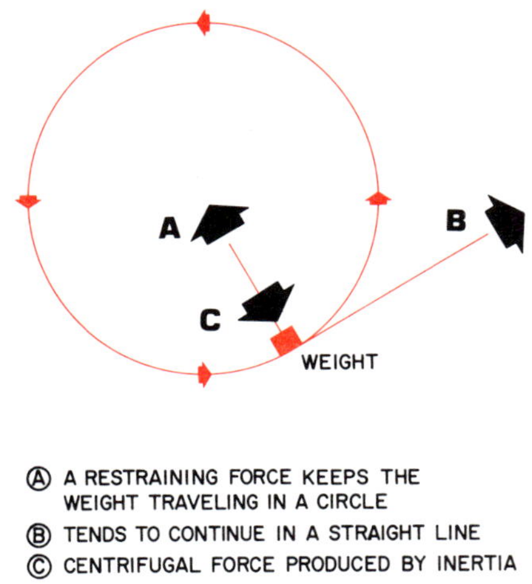

Fig. 2-30. Inertia is a force that tries to keep an object in motion in a straight line. Straight-line force produces a force away from the center. It is called "centrifugal force."

These are of the kinds most often used in home and industry.

Friction Clutch. The friction clutch is the type of clutch used on automobiles with standard transmissions. It has two discs that are brought together under pressure. See Fig. 2-28. In this example, a friction disc of asbestos and other materials is pressed between two metal discs. The friction material on the friction disc grips the metal plates and locks the units together.

Positive Clutch. The positive clutches, shown in Fig. 2-29, will not slip. The clutch must be engaged and disengaged when the clutch parts are standing still or moving at low speeds.

Centrifugal Clutch. Centrifugal clutches work on the principle of centrifugal force. This principle is based on the law of inertia. This law states that a moving object tries to keep going in a straight line. See Fig. 2-30. A force must be applied to keep an object in a circular path. This force increases as the speed of rotation increases.

In the centrifugal clutch, a pair of weights is attached to an input shaft. See Fig. 2-31. The weights move out as the shaft rotates. This movement is produced by the weights as they

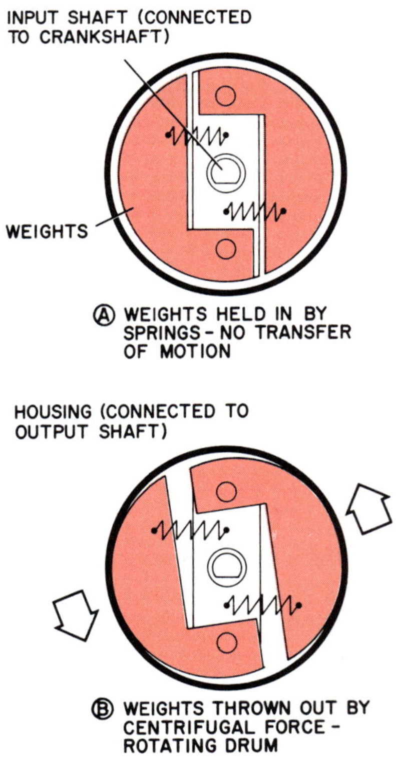

Fig. 2-31. How the centrifugal clutch operates.

A. Centrifugal clutch mounted on engine output shaft.

Ohlsson & Rice, Inc.

B. Disassembled view of centrifugal clutch and reduction gears.

Fig. 2-32. The centrifugal clutch used on a small gas engine.

try to move in a straight line. The weights drive the housing by friction. The outward force produced by the weights increases as the rotation speed increases. A housing surrounds the weights and keeps them from flying away from the shaft.

The housing is connected to the output shaft. The centrifugal force keeps the weights against the housing. Power is transferred from the input shaft, through the weight, to the housing and output shaft.

The centrifugal clutch can rotate in either direction. Also, the force varies with the rotational speed. As speed increases, more centrifugal force is developed. This maintains a positive transfer of power.

Figure 2-32 shows a clutch of this type. View **A,** shows the clutch assembly mounted on a small engine. The engine output shaft (crankshaft) drives the weights. As the weights swing out, they press against the housing. This forces the housing to rotate with the weights. In view **B,** the back of the housing contains a gear. This gear meshes with the large gear. It provides a speed reduction. Speed can also be reduced by meshing a second large gear with the small gear mounted on the first large gear.

These clutches are useful in keeping input and output disengaged until a minimum speed is reached. For example, they can be used on the output of a small engine rotated by a pull cord. Starting speed is too slow to engage the clutch. After starting, the engine speed pro-

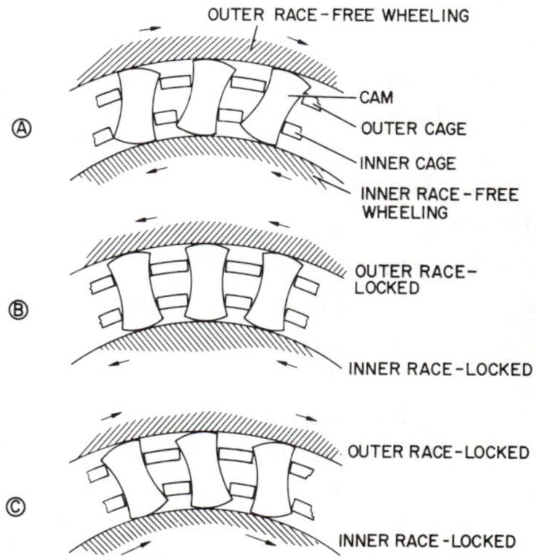

Fig. 2-33. The operation of an overrunning or freewheeling clutch.

Ford Motor Company

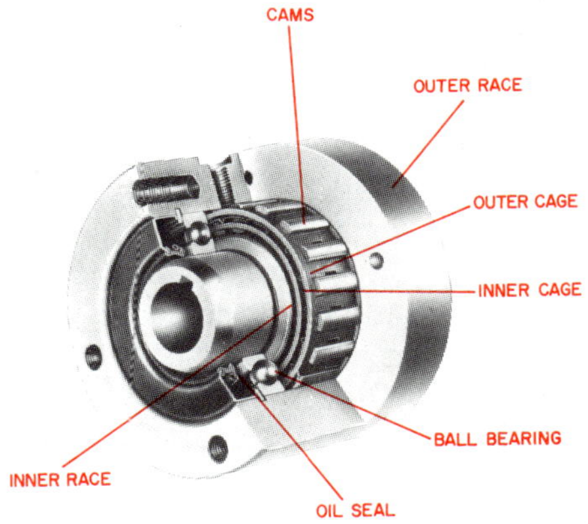

Fig. 2-34. Cutaway view of the overrunning clutch described in Fig. 2-33.

Ford Motor Company

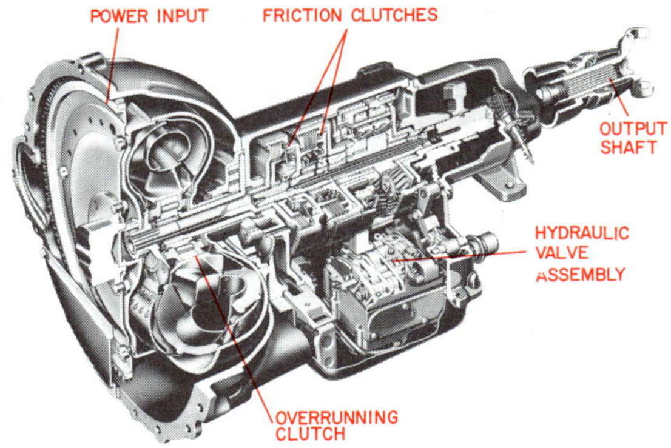

Fig. 2-35. A cutaway view of the modern automatic transmission to show operating units.

duces enough centrifugal force to operate the clutch.

Overrunning or Freewheeling Clutches. These clutches provide a transfer of power in one direction only. Figure 2-33 shows how an overrunning clutch uses cams to transmit power from the input to output shaft.

If the inner and outer races are turned, no power is transmitted. See view **A** of Fig. 2-32. Each race is turning free of the other. This is called **freewheeling.**

In view **B,** input is through the outer race. Each cam is rotated slightly, locking the units together. The outer race drives the inner race. If the inner race rotates faster than the outer race, the cams shift back to the position shown in view **A.** When the inner race rotates faster, it overruns the outer race.

In view **C,** input is in the reverse direction and through the inner race. This diagram shows the cams turned further. They are transmitting a heavy load. In this case, the outer race is driven and could overrun the inner race. If the inner race is stopped, the outer race can keep turning. This is freewheeling.

Automobile automatic transmissions use this type of clutch. It is also used in industry. This might be in cases where the output shaft must rotate freely after the input shaft stops. This was described above. Figure 2-34 shows a typical overrunning clutch. Figure 2-35 shows a

A. Roller chain coupling.

Morse Chain Company

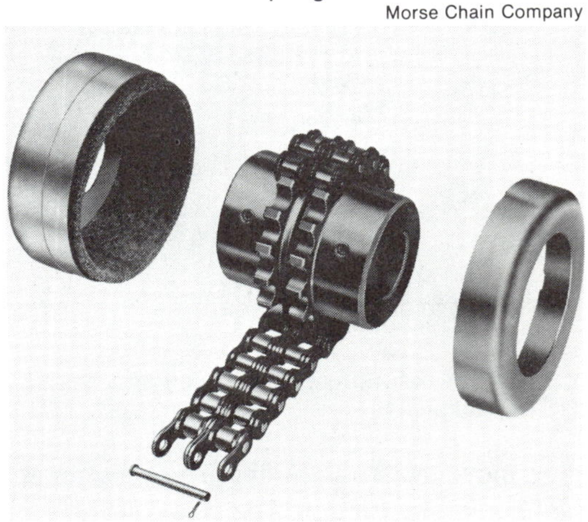

B. Disassembled coupling with protective housing.

Fig. 2-36. Roller chain couplings using two sprockets and chain to transmit power.

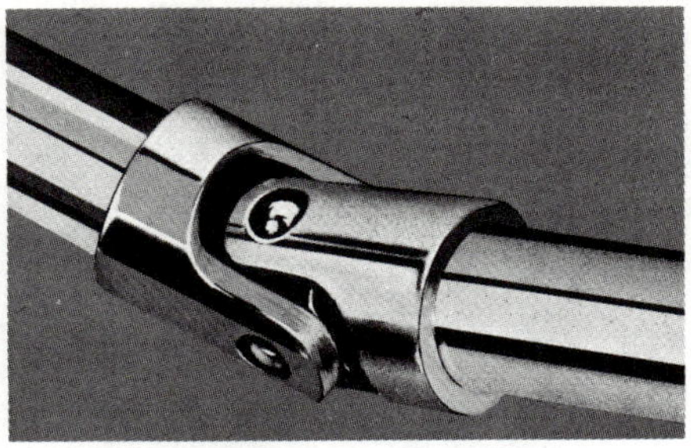

Fig. 2-37. Universal Joint

modern automatic transmission. It contains an overrunning clutch and two friction clutches. Its operation is complex and is left for advanced study.

Couplings

Couplings are permanent connections for the transmission of power. They are placed in the system to provide directional change. They also connect lengths of shaft.

There are many types of industrial couplings in use. Some are rigid. They are used for connecting separate shafts. See Fig. 2-36. All three shown use two-toothed sprockets and a link chain. One sprocket is fastened to the end of each shaft being connected. The chain is then wrapped around both sprockets. This locks them together. Power is transmitted from one shaft to the other through the chain.

The universal joint is a coupling that provides for alignment change. See Fig. 2-37. The universal joint permits the shafts connected to it to move. If a rigid coupling were used, flexible alignment of the shafts would not be possible.

The universal joint shown in Fig. 2-37 has two sets of bearings on the yoke. One shaft is connected to each set of bearings. The power is transmitted from one shaft to the other through the yoke. The bearings between the yoke and shaft permit up and down flexing of each shaft during operation.

TRANSMISSION

The main power transmission devices are the control devices already described. Often, the only transmission needed between the input and output is one or more control devices. These would be for changing the kind of the motion. Motion might be changed from reciprocating to rotary, fast rotation to slow rotation, and so on.

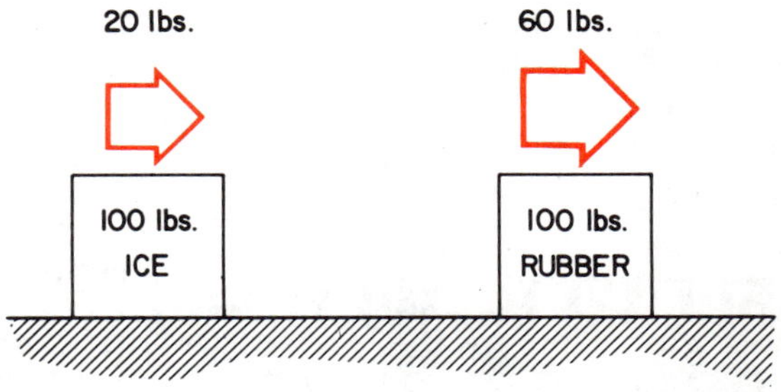

U.S. Customary	SI Metric
20 lbs.	9 kg
60 lbs.	27 kg
100 lbs.	45 kg

Fig. 2-38. Friction is the resistance to motion between two objects. The friction for ice is less than for rubber.

Friction

Whenever a mechanical transmission of power is planned, friction must be considered. There will be friction loss in control devices and along transmission lines. This can greatly reduce the available power.

Whenever a moving surface contacts a stationary surface, the stationary surface slows, or tries to slow, the motion. The slowing action is called **friction.** See Fig. 2-38. Friction even occurs between air and a mechanical device in motion. The air slows down the device. Power is used to overcome the friction.

In mechanical power transmission, friction produces the greatest power loss. The more complex the device, the greater the loss.

POWER OUTPUT

After the motion has been changed and carried to the place where needed, it is put to work. The final use, or output, may be moving an automobile, or it might be cutting the lawn or operating a complex piece of machinery.

FLUID POWER – PRINCIPLES AND THEORY

Chapter 3

Liquids and gases are the fluids of fluid power. The most common fluids used to transmit power are air and oil. In the typical fluid power system, the fluid is placed under pressure. It is carried to the place where power is needed. The pressure on the fluid is changed to motion by a motor or piston.

Using both air and oil to transmit power has developed into major fields of engineering. Engineers refer to air or any other gas used in fluid power as **pneumatics.** They refer to oil or other liquids used to transmit power as **hydraulics.**

Fig. 3-1. Uncontrolled fluid power.
Nature uses air (pneumatics) and water (hydraulics) to transmit power. Fluids can be controlled to provide useful work.

MEASUREMENT OF PRESSURE

Fluid power is the control and transmission of power by the use of a fluid under pressure. Pressure is a measurement of force over a given area. This area is usually one square inch. The basic unit of pressure is pounds per square inch (psi).[1]

Atmospheric pressure is the standard used to measure gas pressures. The air that surrounds the earth applies a force of 14.7 psi on the surface of the earth, Fig. 3-2. This figure is often rounded to 15 psi.[2]

SI Metric Conversion

1. Pressure is measured in pascals (Pa). The pascal is 1 newton per meter squared. It takes 6895 Pa or 6.895 kilopascals (kPa) to equal 1 psi.
2. Atmospheric pressure is equal to about 100,000 Pa or 100 kPa.

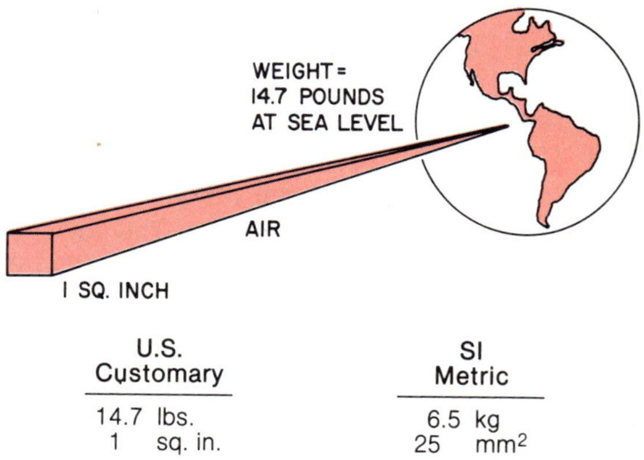

Fig. 3-2. Air surrounding the earth produces an atmospheric pressure of 14.7 psi (101 kilopascals). This pressure is applied to the entire surface of the earth.

On some pressure gauges, the 14.7 psi atmospheric pressure is set to read at zero psi. These gauges must show pressures above or below zero. (This is above or below the 14.7 psi atmospheric pressure.) See Fig. 3-3. Pressures above zero are shown in pounds per square inch. To show that they are based on an atmospheric pressure of zero psi, the word "gauge" is added to the reading. Atmospheric pressure, therefore, is zero pounds per square inch gauge (psig).[3]

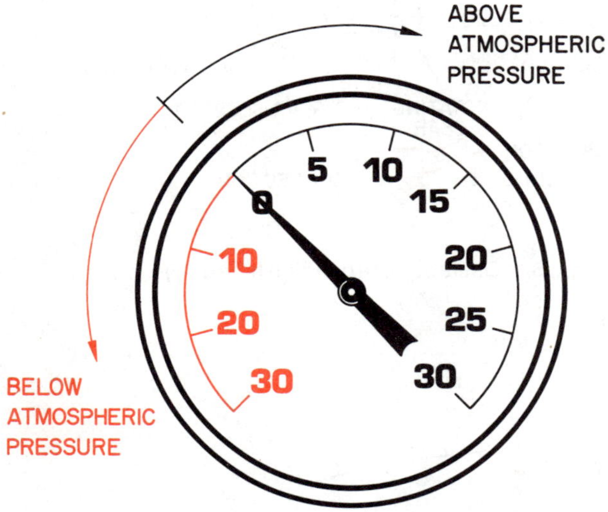

Fig. 3-3. A gauge that shows atmospheric pressure as 0 psi (0 Pa) must show pressures above and below zero.

A pressure below atmospheric pressure is called a **vacuum.** The measurement of a vacuum is based on a method used to show that atmospheric pressure is about 15 psi. Mercury, a liquid metal, weighs about 1/2 pound per cubic inch. As a result, a one square inch column of mercury 30 inches high weighs 15 pounds. See Fig. 3-4. This weight produces a pressure of 15 psi. It should balance atmospheric pressure.[4]

View **A** of Fig. 3-4 shows an open tube standing in a dish of mercury. Atmospheric pressure (15 psi) is equal inside and outside of the tube.

If the tube is sealed at the top and one-third of the air pumped out, the pressure will drop to 10 psi. See view **B** of Fig. 3-4. The atmospheric pressure on the outside will push the mercury up the tube. It will stop moving up

SI Metric Conversion

3. If this same type of pressure gauge were used in metrics, atmospheric pressure would be zero pascals.
4. Vacuum is also measured in pascals. However, Fig. 3-4 should be studied only in English measurements. SI metric uses no mercury measurement for pressure gauges.

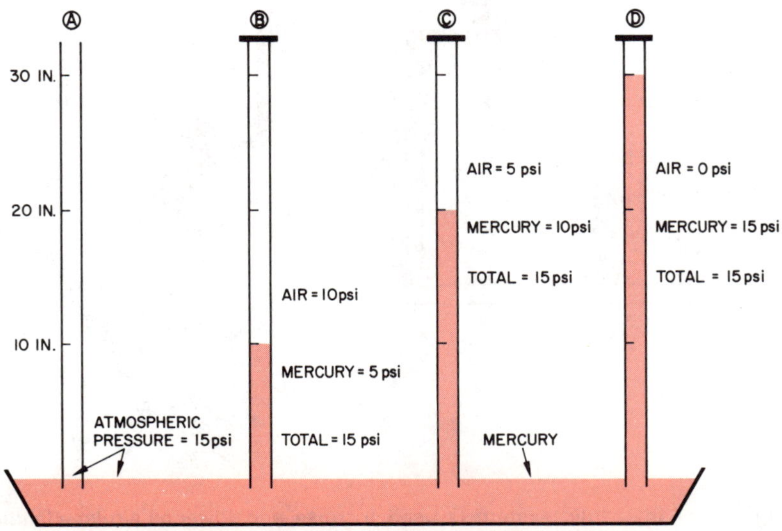

Fig. 3-4. The height of a column of mercury is used as a measurement of vacuum (pressure less than atmospheric).

when the pressure produced by the weight of the mercury, plus the remaining air pressure, balances the atmospheric pressure. The mercury column is 10 inches high. At 1/2 psi for each inch, 10 inches produces 5 psi. Added to the remaining 10 psi of air pressure, the needed 15 psi is produced.

View **C** of Fig. 3-4 shows more removal (5 psi) of air. The mercury now stands at 20 inches, since 10 psi are needed to balance the atmospheric pressure of 15 psi.

View **D** shows all of the air removed. No air pressure remains in the tube. The atmospheric pressure of 15 psi is balanced by an equal pressure produced by the 30-inch column of mercury. This is the highest level a column of mercury can be pushed by atmospheric pressure.

Gauges operate on the same principle. A bourdon tube, Fig. 3-5 is the main part of the gauge. Increased pressure in the tube makes

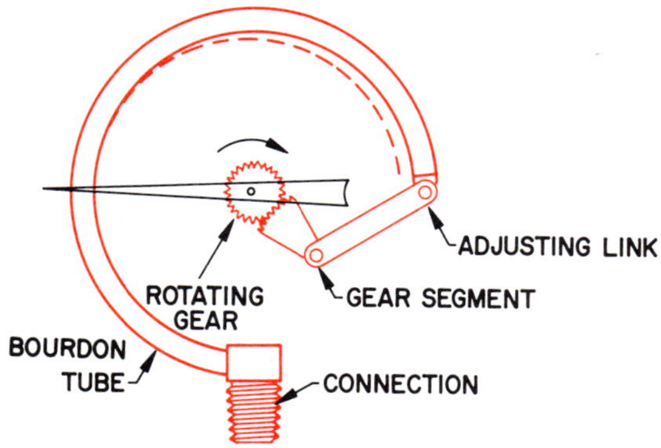

Fig. 3-5. Operation of a bourdon tube gauge.

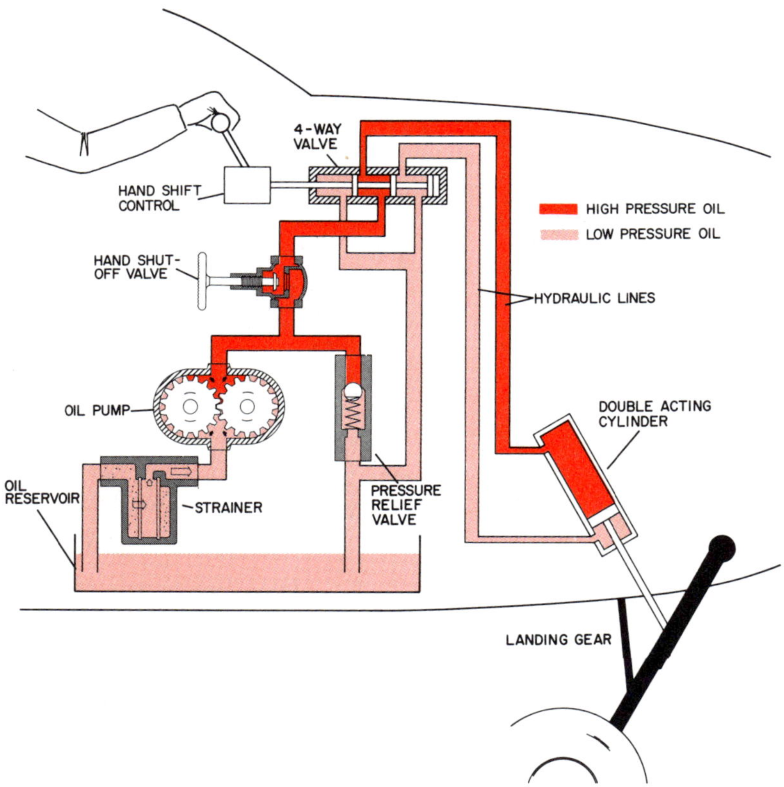

Fig. 3-6. A hydraulic system is used to raise and lower an airplane landing gear. The system is in the gear down position.

the tube straighten a little. This action rotates the gear segment. The gear's action moves the needle to show the increased pressure.

FLUID POWER SYSTEMS

Fluid power systems control and transmit power. They are not sources of power. They use outside sources of power such as an engine or electric motor.

The fluid power system has the same elements that make up the mechanical system. These are input, control, transmission, and output.

Hydraulic System

Figure 3-6 shows the hydraulic system used to raise and lower the landing gear of an airplane. The same kind of hydraulic system is used to raise and lower convertible tops of automobiles. It also raises and lowers car windows or the ram of a large industrial press.

The oil pump moves oil from the reservoir into the high pressure lines. See Fig. 3-6. The oil travels past the hand shut-off valve into the hand operated four-way valve. In the position shown, oil is directed to the upper end of the double-acting cylinder. The high pressure oil extends the piston of the double-acting cylinder. The piston lowers the landing gear. As the piston extends, oil in the underside (rod end) of the cylinder is forced into the low pressure line. The oil travels through the four-way valve. It then returns to the reservoir. The pressure relief valve protects the system. If the pressure becomes too high, the pressure relief valve opens. This lets the high-pressure oil return to the reservoir.

To raise the landing gear, the hand-operated control valve is shifted (pushed back). See Fig. 3-7. The high pressure oil is now directed to the rod end of the double-acting cylinder. This action retracts the piston. This in turn causes the landing gear to raise. The shifting

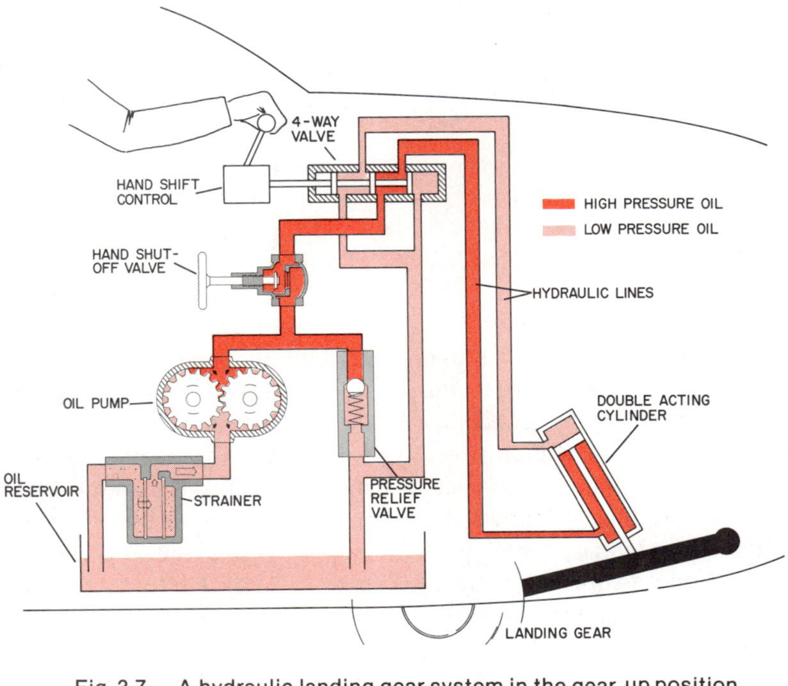

Fig. 3-7. A hydraulic landing gear system in the gear up position.

Alfred A. DeLardi, F.P.S.A.

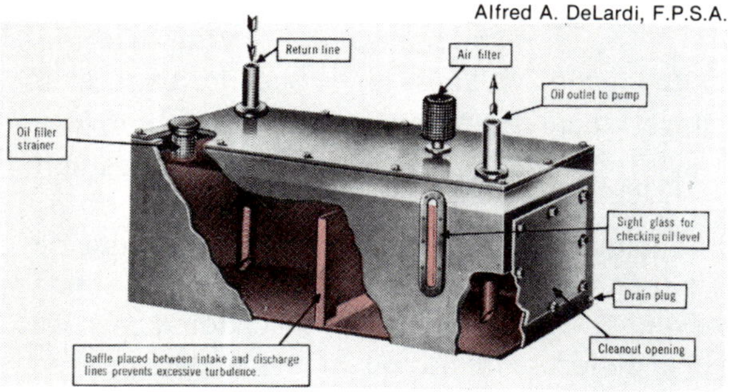

Fig. 3-8. Typical oil reservoir.

of the valve also opens the upper end of the cylinder to the reservoir. This permits the return of oil being forced out of the cylinder.

Reservoir, Hydraulic Fluid, and Filters

The reservoir is not an operating part of the system. The reservoir serves as the storage unit for oil not in use. See Fig. 3-6. The pump takes oil from the reservoir. The output cylinder or motor returns the oil after use. The reservoir is usually kept at atmospheric pressure.

The reservoir shown in Fig. 3-8 contains an air filter. The filter keeps dirt from getting in the system. The level of the oil may go up and down during operation. This change causes a small movement of air to pass in and out of the reservoir.

Hydraulic Pumps

The hydraulic pump is often called the heart of the hydraulic system. The pump provides the system with fluid under pressure. It is like the heart supplying the body with blood under pressure. Fluid under pressure and in motion is power. The pump changes mechanical motion from an engine or electric motor into fluid motion. Or it may change mechanical power into fluid power. Figure 3-6 shows the location of the oil pump in a fluid power system.

There are many pump designs. A reciprocating pump discussed here shows the basic principles of pump operation. The most common of the rotary pumps are the gear pump and the vane pump. The centrifugal pump will show the operation of a special type of pump. It is called a non-positive displacement pump.

A **positive displacement** pump is one in which a fixed amount of fluid is entrapped. The fluid is moved from the inlet to the outlet side of the pump. The low pressure produced on the inlet side is filled with oil from the reservoir. This action is called **charging** the pump. The gear, vane, and reciprocating piston pumps are all positive displacement pumps.

A **non-positive displacement** pump is one that pumps a varying amount of fluid. The fluid is moved from the inlet to the outlet by an impeller. The impeller increases the speed of the fluid as it passes through the pump. The fluid is not entrapped. It is only caused to move faster. As a result, this pump is usually charged by gravity.

Reciprocating Pump. Hydraulic pumps take oil from the reservoir tank and force it into the system. Oil will not compress. When the pump puts oil into the system, motion must take place at the output side of the system.

The simple piston type pump is shown in Fig. 3-9. On intake, as shown in view **A,** the piston moves down. This action creates a vacuum in the piston chamber. Atmospheric pressure within the reservoir opens the intake valve. At the same time, it pushes the oil up into the vacuum of the piston chamber. During this action, the high pressure oil in the discharge line tries to move back into the chamber. But the oil pushes against the backside of the discharge valve. It pushes the valve tightly against the opening. This action closes the valve and prevents the discharge oil from entering the chamber.

When the chamber is filled with oil from the reservoir, the piston reverses direction. As the piston moves up, oil is forced out of the chamber. See view **B** of Fig. 3-9. First, the oil tries to go back into the reservoir. However, the intake valve is pushed against its seat as the oil tries to flow into the reservoir. The discharge valve is opened. The oil moves into the

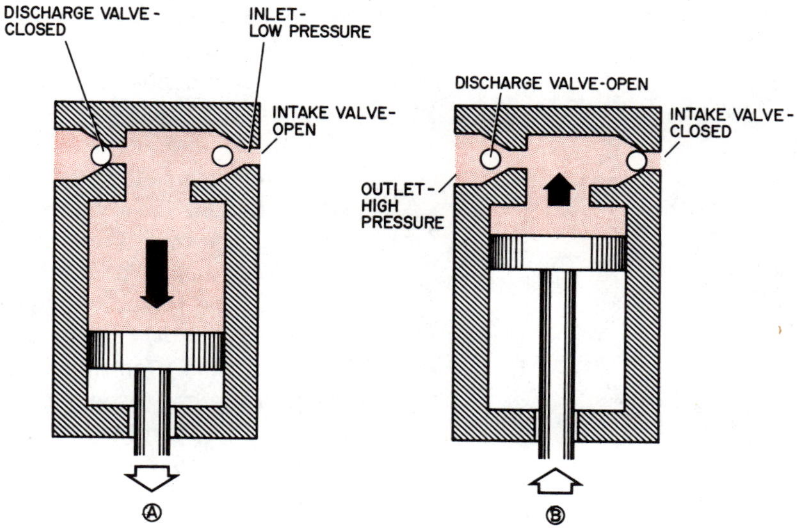

Fig. 3-9. Operation of a simple piston type of pump.
A. Intake stroke — piston moves down.
B. Discharge stroke — piston moves up.

discharge line. The upward movement of the piston pushes most of the oil from the chamber into the discharge line.

After discharge, the piston moves down for another intake stroke. More oil is taken from the reservoir. The cycle is repeated.

This pump is called a reciprocating pump. It is driven by an engine or electric motor that provides rotary motion. A crankshaft and connecting rod system is part of this pump. See Fig. 2-5. It changes rotary motion into reciprocating motion.

Gear Pumps. The gear pump operates on a simple principle, Fig. 3-10. The pump consists of two gears. One gear is driven by the input source of power (engine or motor). This gear is in mesh with the other gear. It rotates that gear in the opposite direction. The oil on the inlet side of the pump is picked up by the gears as they pass the inlet port. The entrapped oil is carried between the gear teeth as they move along the outer edge of the pump. It is forced into the outlet (discharge) side of the pump. The oil cannot return to the inlet since the gear teeth mesh, forcing out the oil.

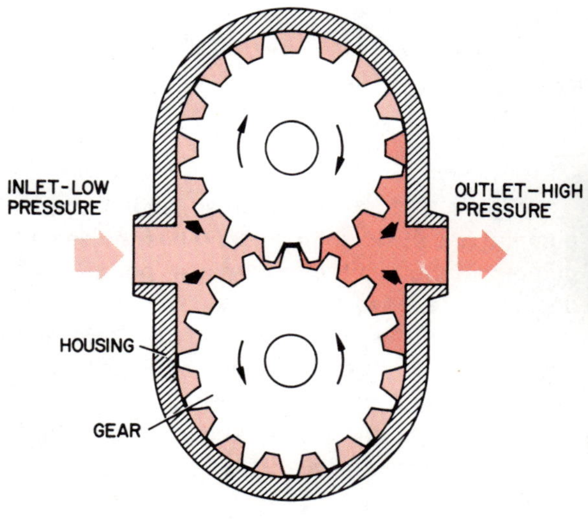

Fig. 3-10. Operation of a gear pump.

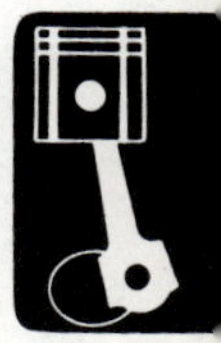

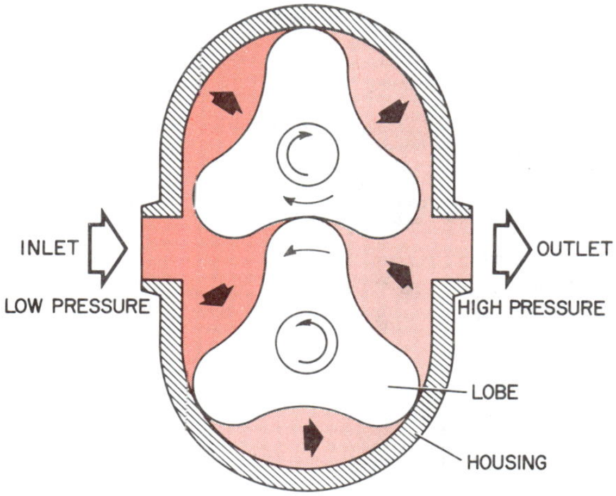

Fig. 3-11. Operation of a lobe (rotor) pump.

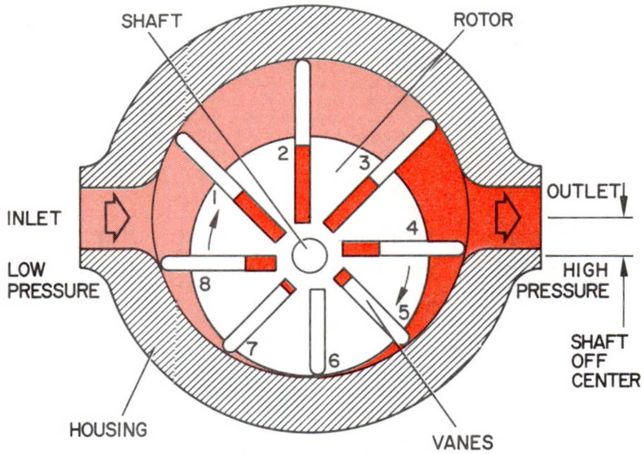

Fig. 3-12. Operation of a vane pump.

Figure 3-11 shows another kind of gear pump. It is called a **lobe pump.** It has the same pumping action as the regular gear pump. The design of the lobes causes this device to pump more oil than the regular gear pump. However, one lobe cannot drive the other lobe. Both lobes must be driven.

Vane Pump. The vane pump uses a series of sliding vanes to move oil from the inlet to the outlet side of the pump. See Fig 3-12. The vane pump has a rotor, group of sliding vanes, drive shaft, and housing. The drive shaft rotates the rotor. The rotor is not centered in the pump housing. It is centered towards one side or the bottom of the housing. This position provides a large space on one side of the rotor.

In operation, the rotor moves in a clockwise direction. The vanes are pushed from the center of the rotor. This causes them to ride against the inside of the pump housing. The vanes may be forced out by small springs. Or they may be thrown out by centrifugal force as the rotor turns. Centrifugal force is the most common method used. See Fig. 3-12. During operation, oil from the inlet side (low pressure) is permitted to flow behind the vanes. This helps to hold the vanes out against the housing. The rotor is drilled to permit this movement of high pressure oil.

As the vanes rotate, oil is trapped between two vanes (1 and 2) on the inlet side. As the rotor rotates, the vanes will carry this oil to vanes 3 and 4, the discharge side of the pump. As the vanes move into this position and continue on to position 4 and 5, the oil will be forced from the chamber and into the discharge line. Vanes 1 and 2 will continue around and repeat the cycle.

Note that each pair of vanes, 1 and 2, 2 and 3, 3 and 4, etc., pump oil from inlet to outlet.

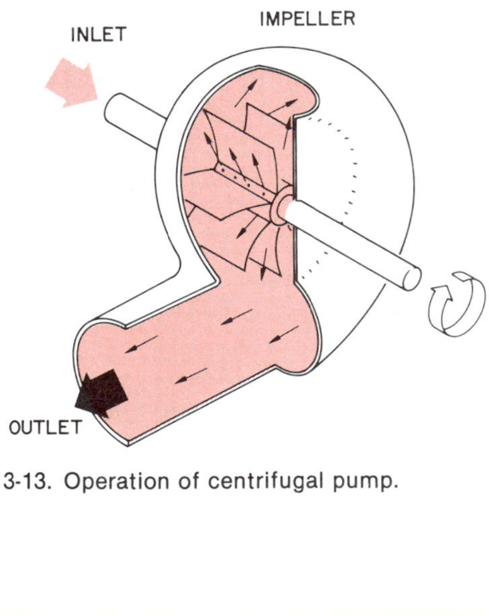

Fig. 3-13. Operation of centrifugal pump.

Chrysler Corp.

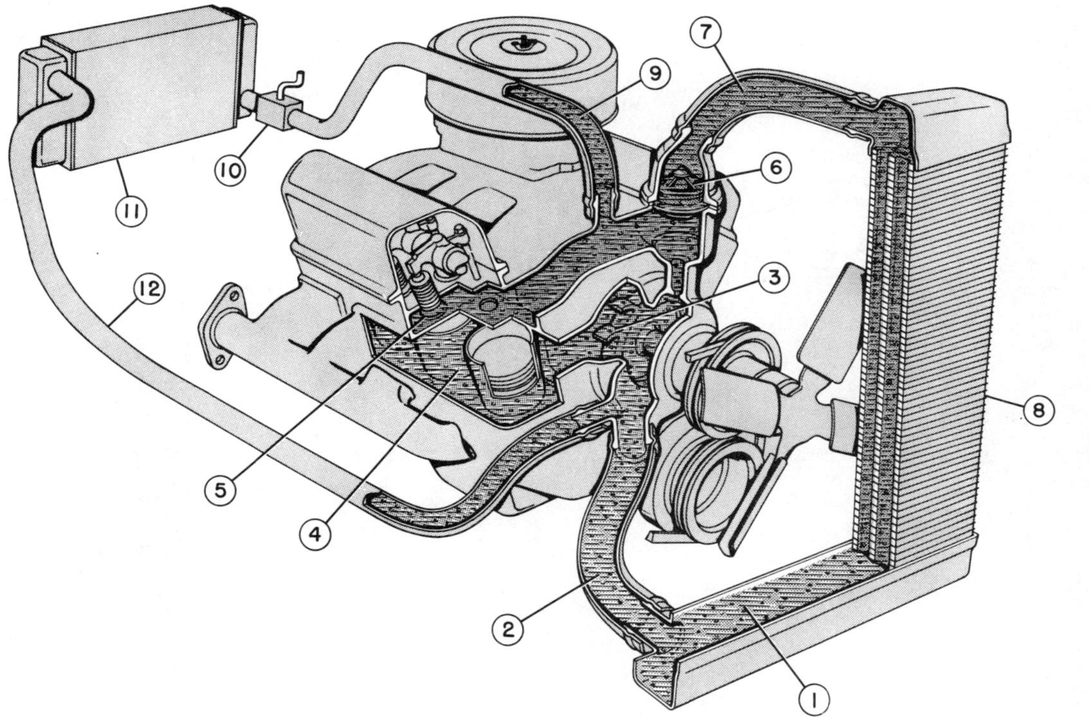

1. COOLANT IN RADIATOR
2. COOLANT BEING DRAWN BY PUMP
3. WATER PUMP
4. COOLANT CIRCULATING AROUND CYLINDERS
5. COOLANT CIRCULATING AROUND VALVES
6. THERMOSTAT
7. HEATED COOLANT RETURNING TO RADIATOR
8. RADIATOR
9. HEATER COOLANT HOSE-TO HEATER
10. WATER SHUT-OFF VALVE
11. HEATER CORE
12. HEATER RETURN HOSE

Fig. 3-14. Water flows through an automotive engine and heater.

This action produces a continuous flow of oil into the discharge line.

Centrifugal Pump. Most non-positive displacement pumps operate by centrifugal force. See Fig. 3-13. The fluid enters from the reservoir through the center of the pump. The impeller blades start the fluid revolving. This increases the speed of the fluid. As the fluid revolves, centrifugal force moves it out to the inner wall of the housing.

A centrifugal pump is commonly used in the auto engine's cooling system. See Fig. 3-14. Water is pumped through the engine to remove heat not needed by the engine. When the engine is cold, a valve called a **thermostat** closes. It stops the flow of water until the engine has warmed up.

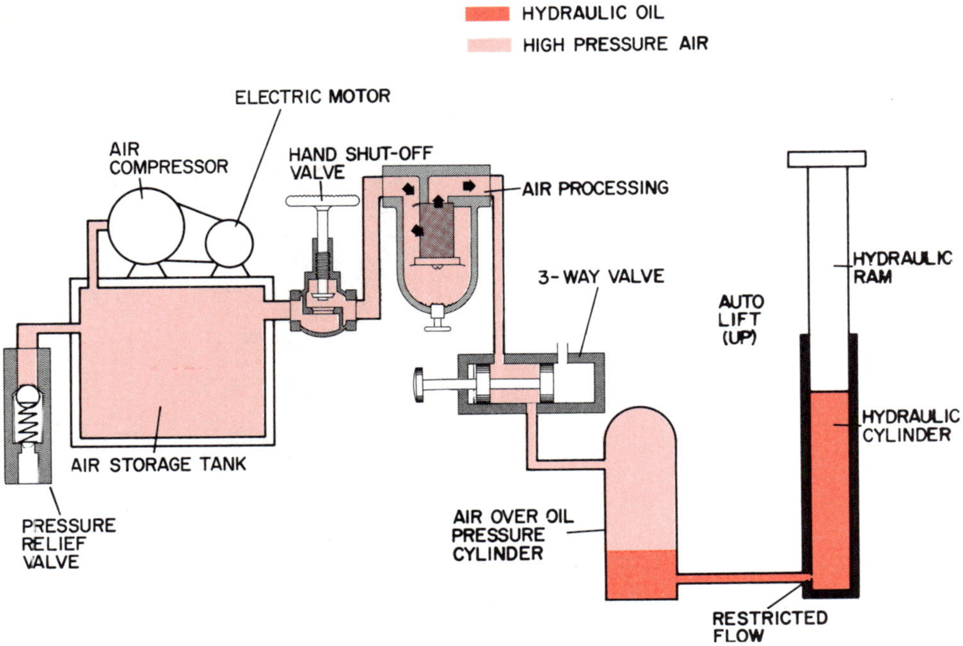

Fig. 3-15. The pneumatic system is used to raise and lower an automobile hoist. This pneumatic system uses oil to raise the hoist. For this reason, it is called an "air over oil system."

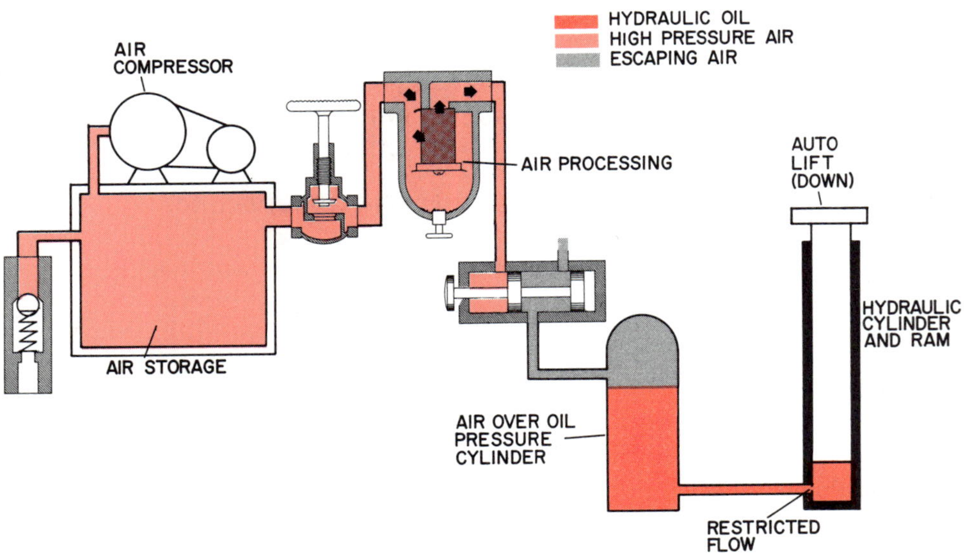

Fig. 3-16. Automobile hoist in a lowered position.

When the three-way valve is opened to reduce the air pressure, the weight of the ram forces the oil back into the pressure cylinder which lowers the ram, as shown.

Pneumatic System

The pneumatic system, Fig. 3-15, is used to lift cars for lubrication and repair. It uses both air and oil. This system is often called an air-over-oil system. The source of pressurized air is an air compressor. The compressor is driven by an electric motor. The air is stored in a large tank until used. A pressure relief valve protects the system from too much pressure. The air passes through a hand valve and an air processing unit. This unit cleans the air. The high pressure air travels through the three-way valve into the air-over-oil pressure cylinder. The air pressure forces the oil into the hydraulic cylinder. This action raises the ram. (The auto lift is on top of the ram.)

Figure 3-16 shows how the lift is lowered. The three-way valve is shifted. This permits the air over the oil to escape. With the pressure released from above the oil, the weight of the lift and the ram push the hydraulic oil from the cylinder. The oil must pass through a small opening. This opening controls the rate at which the lift is lowered. This is a safety measure which prevents the hoist from coming down too fast. Even if the line breaks, the hoist will come down slowly. Anyone working under the hoist and car would have time to escape.

Air Compressors

The air compressor is the pump of the pneumatic system. It operates on the same principles as the hydraulic pump. The air compressor pumps air instead of oil. Figure 3-16 shows the location of the air compressor in the pneumatic system.

Conditioning Air for Pneumatic Systems

Pneumatic systems use the air in the room in which the system is operating. Often this air contains moisture and dirt. In order to protect the system, the air must be conditioned before it is used. Conditioning of the air takes place after the air is compressed. See Fig 3-16. Conditioning is made up of three operations:

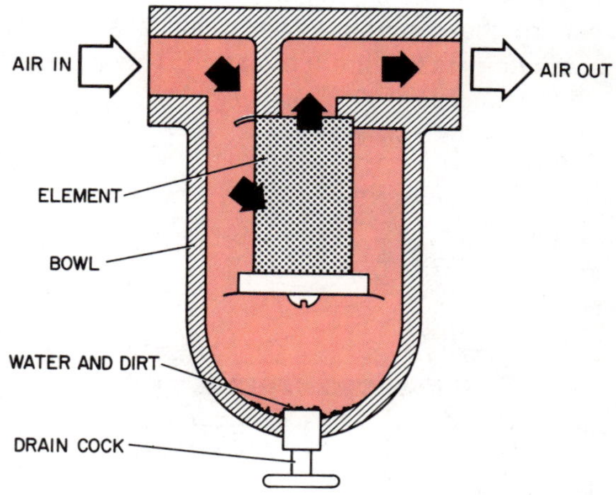

Fig. 3-17. Operation of an air filter containing a manual drain.

1. Air is filtered to remove dirt and moisture.
2. Air is regulated to the proper pressure.
3. A lubricant (usually oil) is added.

Filtering. The compressed air first passes through a filter. See Fig. 3-17. As the air enters the filter chamber, it is directed against the outer surface of the filter. This action produces a swirling motion. Large particles of dirt and water are deposited on the outer surface by centrifugal force.

The air then passes through the filter. The filter is a very fine screen. It takes out the remaining particles of dirt and moisture. Any moisture in the air as a gas or water vapor will pass through this simple filter. If this remaining atmospheric water will damage the system, special drying devices are used. The drain cock at the bottom of the filter should be opened regularly. Water that has condensed in the filter can then drain.

Pressure Regulation. The pressure of a pneumatic system must be controlled. It must provide air at a constant pressure to the system. Air from the filter passes through a regulating valve. See Fig. 3-18. The diaphragm controls movement of air past the valve.

Lubricator. A lubricator is an important part of the conditioning system. Most valves and cylinders require lubrication. This assures good operation and long life. If the air contains a small amount of oil, all parts of the system will be lubricated. This oil is added to the system by the lubricator. Figure 3-19 shows a drip type of lubricator.

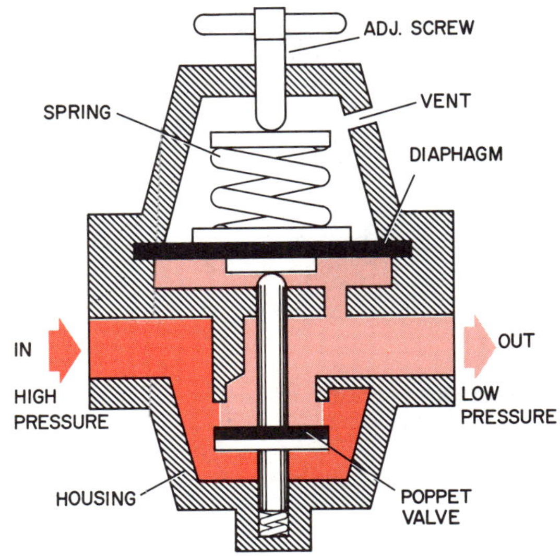

A. Regulator adjusted for slight pressure reduction.

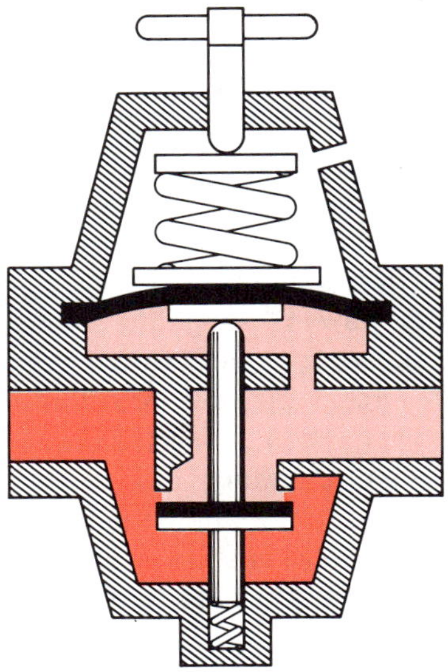

B. Regulator adjusted for major pressure reduction.

Fig. 3-18. Operation of a pressure regulator.

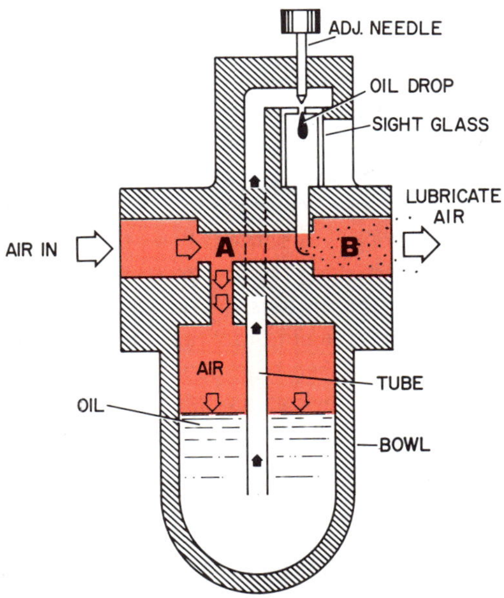

Fig. 3-19. Operation of a drip type of air lubricator.

Sun Oil Company

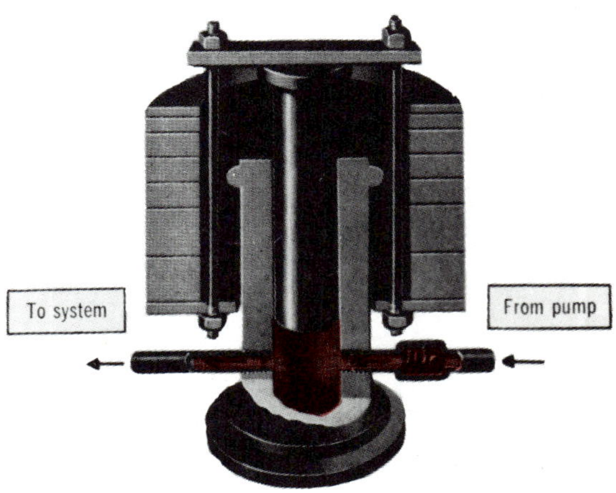

Fig. 3-20. Weight-loaded hydraulic accumulator.

Accumulators and Compressed Air Tanks

On some systems the need for power varies widely. For example, the hoist used in car repair shops needs much power to lift a car into the air. The pneumatic system stores energy during the period between car lifts.

Since pneumatic systems operate under pressure, the storage device must maintain this pressure. Compressed air is pumped into a large tank where it is stored until needed. Condensed water from the air is drained through a valve at the bottom of the tank.

Hydraulic systems need a more complex device. It is called an **accumulator.** One system, a weight-loaded accumulator, is shown in Fig. 3-20. Oil from the pump is stored in the accumulator. As the oil under pressure is pumped into the accumulator cylinder, the piston is lifted. The piston carries the weights up with it. The weights supply the force needed to keep pressure on the oil. When power is needed, the weights push the oil into the system. The pressure is determined by weights and the size of the piston.

An air type of hydraulic accumulator is shown in Fig. 3-21. Compressed air is placed in the tank through the air inlet valve. The compressed air maintains the needed pressure. Oil supplied from the tank is always supplied at the pressure produced by the compressed air. The gauge indicates the pressure reading.

There are other types of hydraulic accumulators. All must have a place for the oil to be stored under pressure. Because oil cannot be compressed, an accumulator is needed for an outside source of pressure.

Input Symbols

Up to this point, units of the power input phase of the system have been shown with a drawing of the part. This procedure is time-consuming. Engineers have developed a "fluid power shorthand." They use symbols to show different parts of the system. These symbols permit a system to be drawn quickly. They also identify what each part means. See Fig. 3-22.

Sun Oil Company

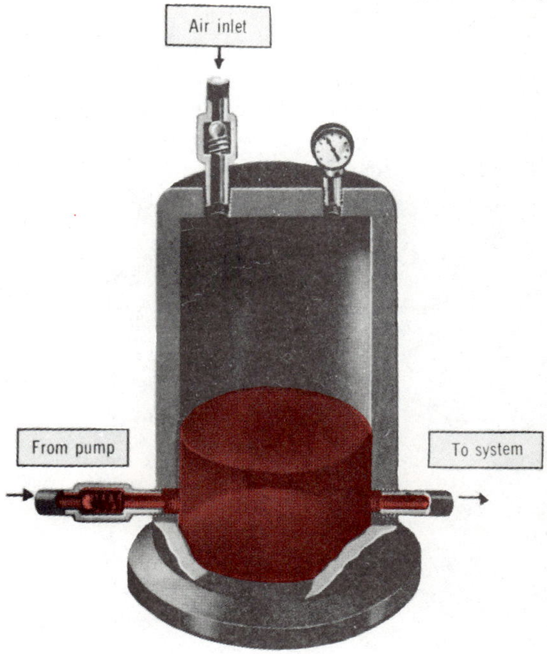

Fig. 3-21. Compressed-air type of hydraulic accumulator.

CONTROL AND TRANSMISSION

Fluid control devices have the same control functions as the mechanical control devices. They can multiply force, change direction of force, and turn power on or off. Fluid controls also adjust and maintain pressure within the system. This supplies fluid in the correct amount and pressure to the system.

Control devices may be located anywhere within the system. The air pressure regulator is located close to the place where the air is to be used. Cylinders that multiply force are located at the point where fluid power is changed back to mechanical power. As a result, control is an important part of the fluid power system.

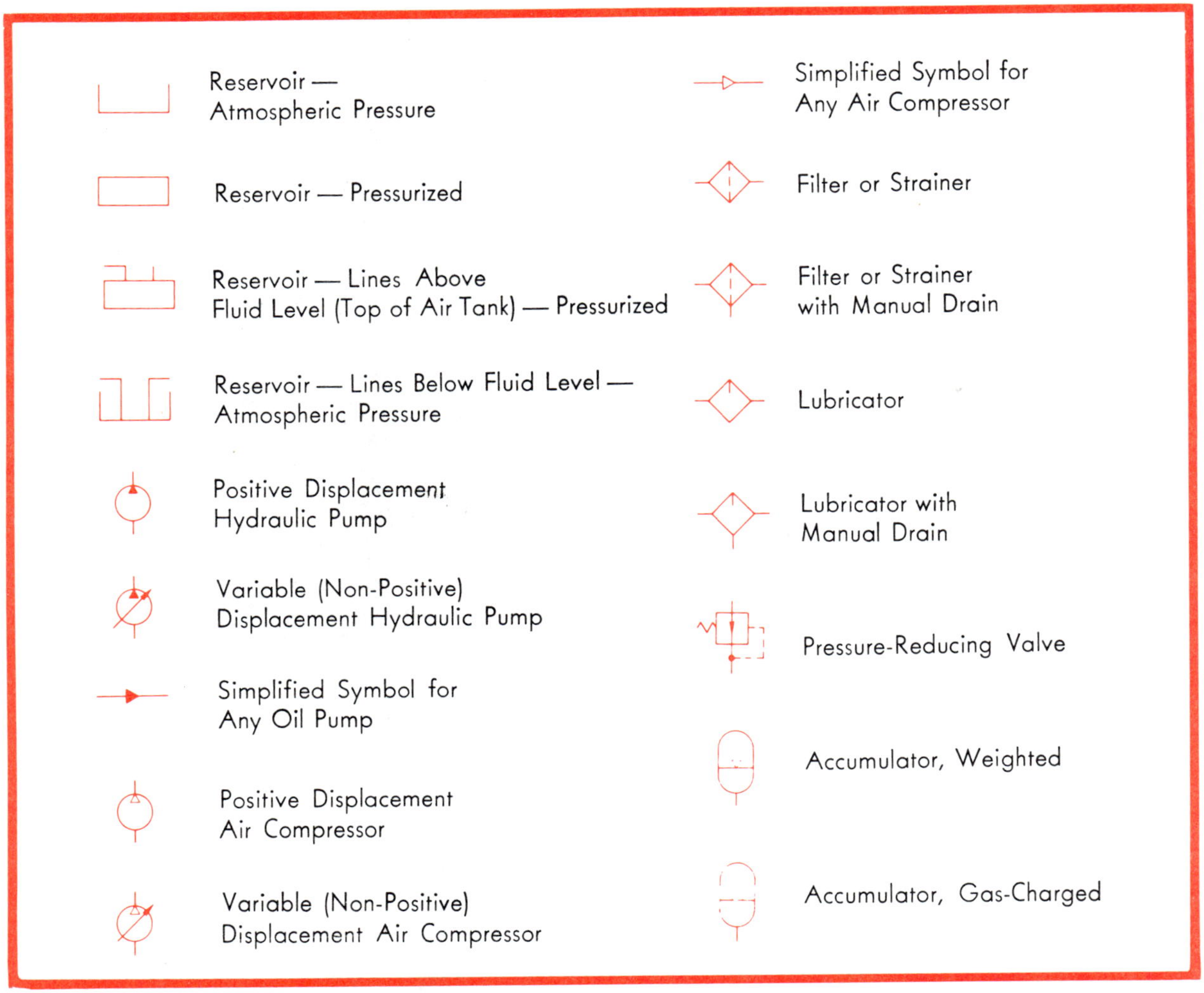

Fig. 3-22. Fluid power symbols for reservoirs, motors, fluid conditioners, and accumulators.

DIRECTIONAL CONTROL VALVES

Directional control valves are usually used to start, stop, or reverse the direction of the flow of the fluid. The most common types of direction flow valves are two-way, three-way, and four-way valves.

Two-Way Valves. Two-way valves have two working connections: (1) input and (2) output. They are used for on-off control of the flow of fluid in the circuit. The valve used to turn on water in a sink, wash basin, or hose is a simple two-way valve. Note the five two-way valves listed here:

1. Globe valve, Figure 3-23.
2. Gate valve, Figure 3-24.
3. Plug valve, Figure 3-25.
4. Needle valve, Figure 3-26.
5. Spool valve, Figure 3-27.

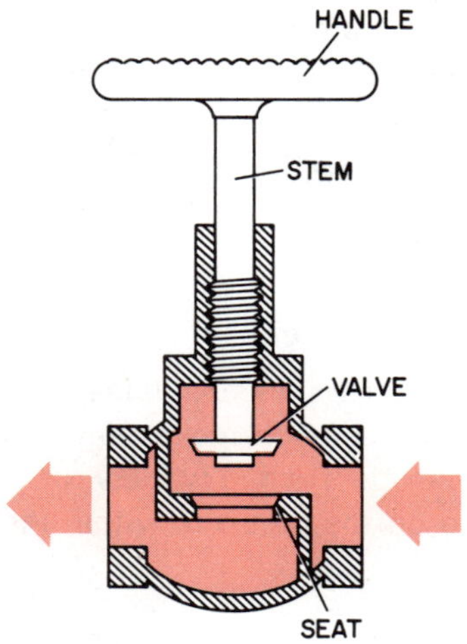

Fig. 3-23. Globe valve in an open position.

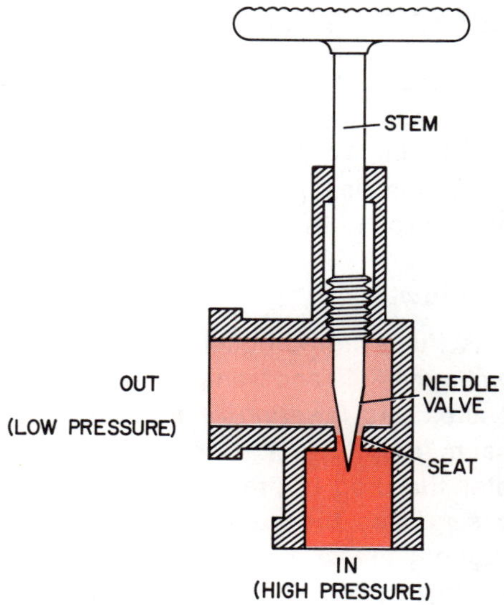

Fig. 3-26. Needle valve reduces pressure as fluid passes through it.

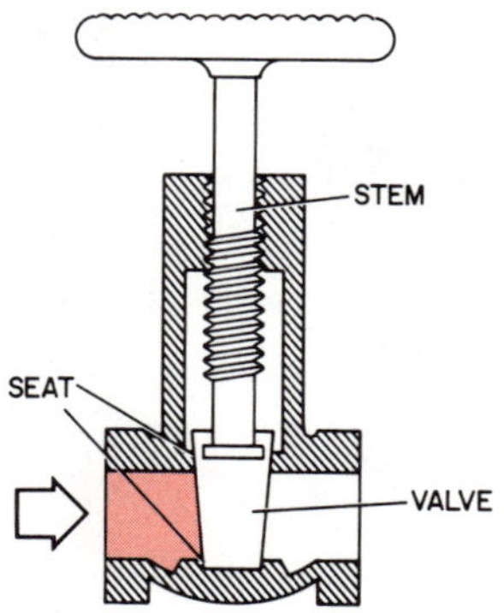

Fig. 3-24. Gate valve in a closed position.

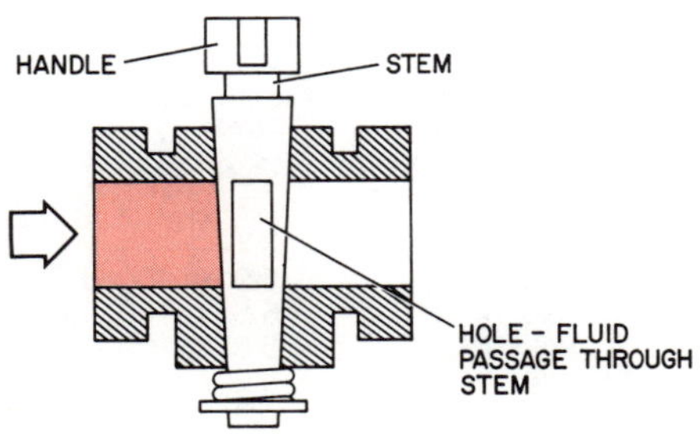

Fig. 3-25. Plug valve in a closed position.

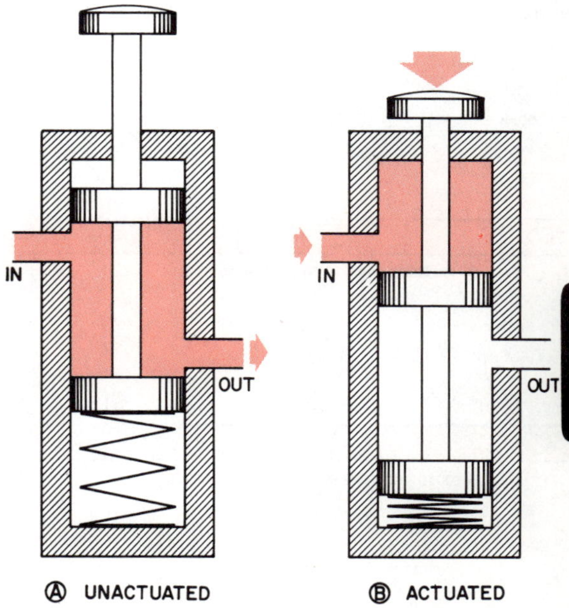

Fig. 3-27. Two-way spool valve with a normally open position.

TRANSMISSION

Oil and air are easily moved from place to place. Transmission lines may be rigid pipes or flexible tubing. With proper strength and size of transmission lines, there will be no leaks or pressure drop in the system.

Fluid Flow

As fluids flow through pipes, the friction between the fluid and walls of the pipe reduces pressure. This lowering of pressure is called **pressure drop.** The longer the distance, the greater the drop in pressure, Fig. 3-28. If the flow is stopped, the pressure will become equal throughout the pipe. See view **B** of Fig. 3-28. Pressure drop occurs only when fluid is flowing.

There are a number of causes of excessive pressure drops in a system. A quick change in direction produces turbulence, Fig. 3-29. Turbulence is a disrupted flow. It causes a big pressure drop in the system.

A restriction can also cause a major pressure drop. Most control valves force the fluid to change direction. As a result, most control valves produce a pressure drop.

Lines that are too small will also produce too much drop in pressure. The faster fluids must flow through a pipe, the greater the friction and loss of pressure. Figure 3-30 shows some typical pressure drops that could occur in a fluid power line.

Transmission Lines

Flexible transmission lines are used on units that have much movement or vibration. They are made from a variety of materials, depending on the pressure within the system.

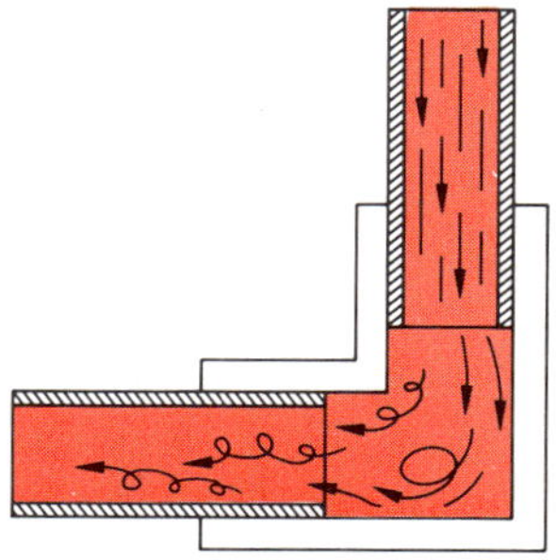

Fig. 3-29. Turbulence results from abrupt change in direction.

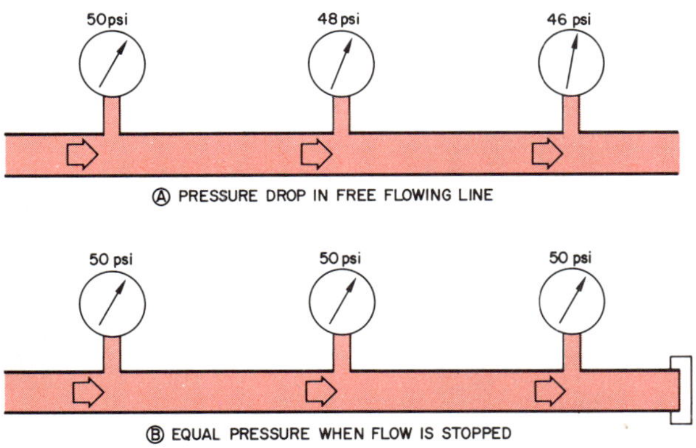

U.S. Customary	SI Metric
50 psi	3.4 × 100 kPa
48 psi	3.3 × 100 kPa
46 psi	3.2 × 100 kPa

Fig. 3-28. Pressure drop in a fluid transmission line.

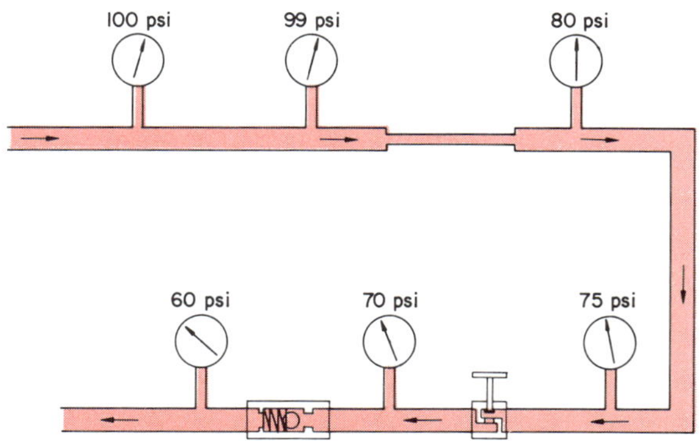

U.S. Customary	SI Metric
100 psi	6.9 × 100 kPa
99 psi	6.8 × 100 kPa
80 psi	5.5 × 100 kPa
75 psi	5.2 × 100 kPa
70 psi	4.8 × 100 kPa
60 psi	4.1 × 100 kPa

Fig. 3-30. Pressure drop in a fluid power system.

Rigid lines are used where flexible ones are not needed. They cost less and give long wear. They may be made of copper, steel, aluminum, or a synthetic material.

The connection of the line to the parts of the system can cause trouble in the fluid power system. Most leaks occur at these connections.

The input and output ports of fluid power units are usually drilled. Threads are cut in the drilled hole. The threads may be either pipe or regular threads. Note the pipe threads in Fig. 3-31. If straight threads are cut, a rubber or synthetic ring (O-ring) is needed. It is placed so that it is pressed tightly against the fitting and housing. This seals the joint. See Fig. 3-32.

Rigid lines use a flared or compression connection. In the flared connection, the tube is flared at the end with a special tool. See Fig. 3-33.

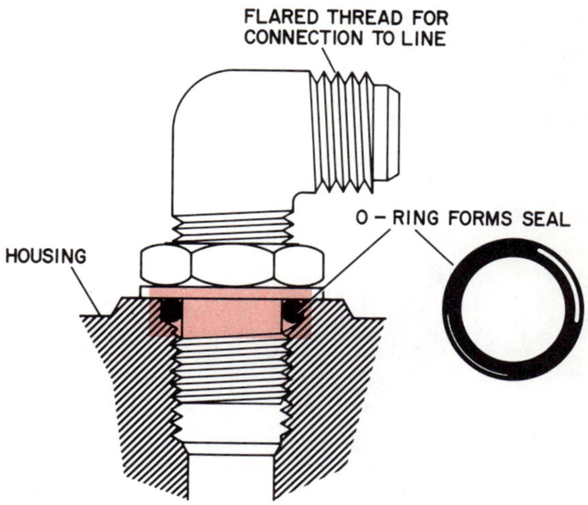

Fig. 3-32. An O-ring is made of rubber or plastic material. It is used to seal straight threads. The seal between the fitting and housing takes place in the colored area.

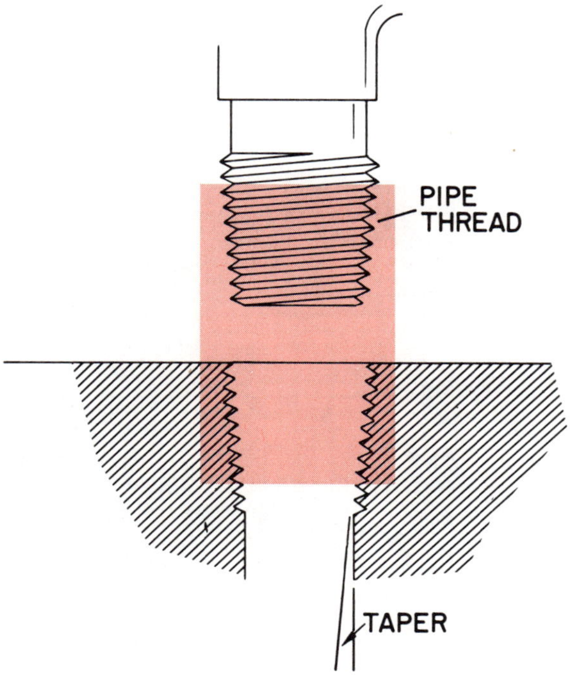

Fig. 3-31. Pipe threads are cut at an angle (tapered). When tightened, the threads come together and form a tight, leak-free joint.

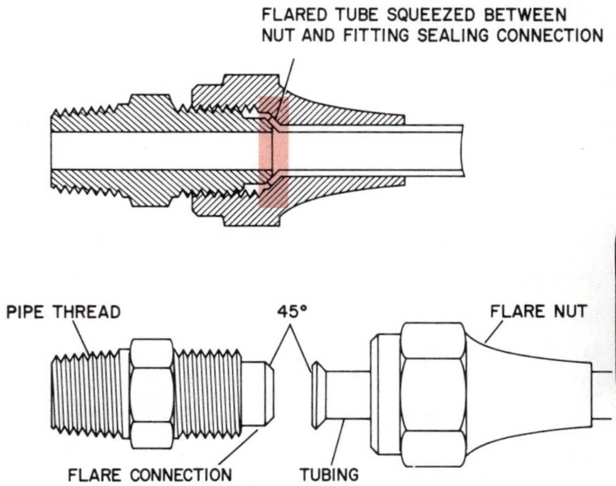

Fig. 3-33. Flared tubing connection. It is used for copper, brass, aluminum, plastic and soft steel tubing. The seal between the line and fittings takes place in the colored area.

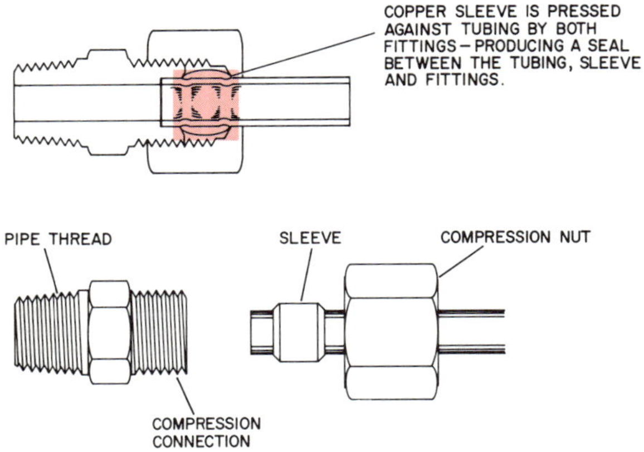

Fig. 3-34. Compression connection.
It is used for copper, brass, aluminum, and plastic tubing. The seal between the line and fittings takes place in the colored area.

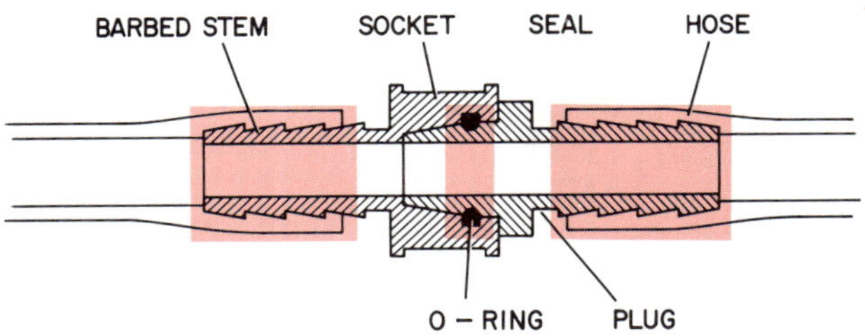

Fig. 3-35. Quick-disconnect coupling.
This is used to connect two pieces of flexible tubing, Colored areas identify places where sealing takes place.

The compression connection has a similar arrangement. It has a copper sleeve instead of the flared tubing. The sleeve is squeezed between the nut and connector. This seals the joint. See Fig. 3-34.

Flexible hose must have a different type of connector to seal the hose to a fitting, Fig 3-35. A barbed stem is inserted into the hose. Since the stem is larger than the hose, the hose tends to mold itself around the stem. This makes a seal and holds the hose on the stem. The other end of the stem contains a special plug designed to snap into a socket. See Fig. 3-35. An O-ring provides the seal between the plug and socket. This unit is called a **coupling** or a **quick-disconnect coupling.**

POWER OUTPUT

Up to this point, power input, control and transmission have been studied. Now you should know how fluid power is put to work for home and industrial use.

The air or oil under pressure is changed from fluid motion to mechanical motion through the use of an **actuator.** An actuator is a device that receives power in fluid form and changes it to mechanical motion. Actuators can produce either linear motion or circular motion. Repeated linear motion is produced by cylinders. It is called **reciprocating motion.** Constant circular motion is produced by fluid motors. It is called **rotary motion.**

Cylinders

Figure 3-36 shows a typical cylinder. When fluid enters the right port (opening) under

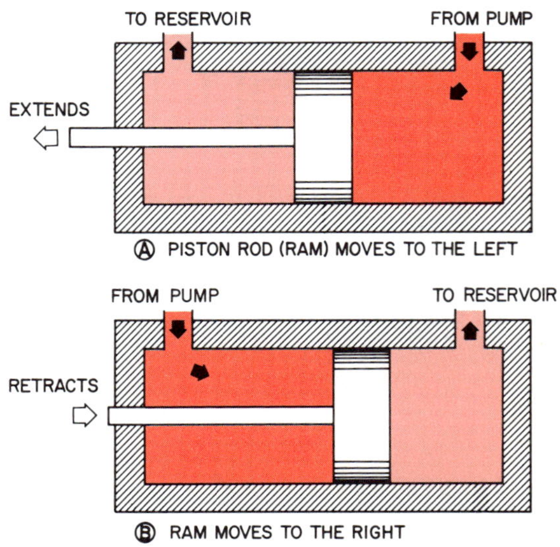

Fig. 3-36. Differential type of double-acting cylinder.

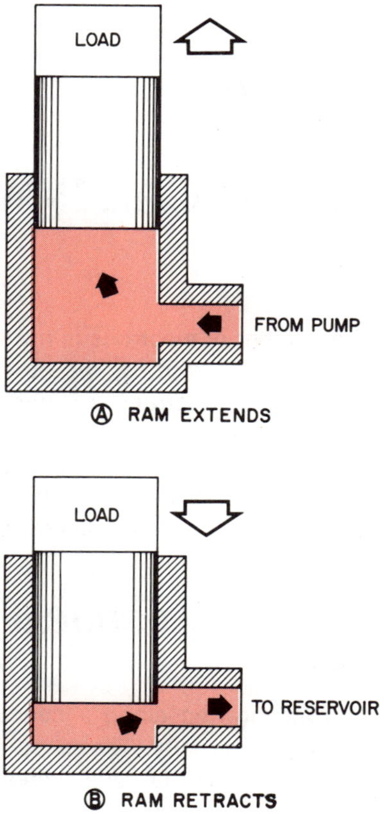

Fig. 3-37. Single-acting cylinder used as a hydraulic jack.

pressure, the piston will move to the left. The fluid trapped in the left end of the cylinder must be removed. If this is an air cylinder, the trapped air can be released to the room. If it is an oil cylinder, the oil must be piped back to the reservoir.

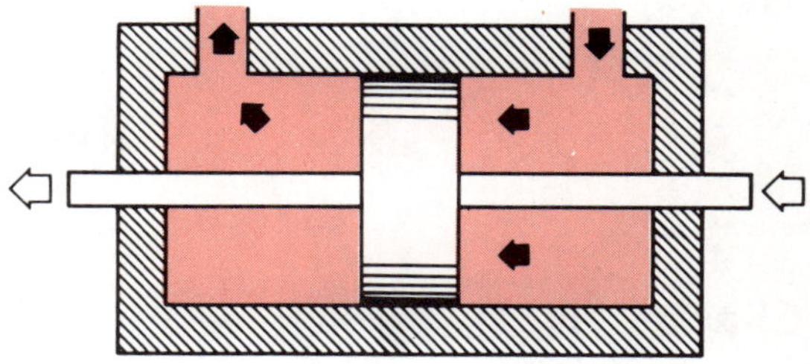

Fig. 3-38. Non-differential type of double-acting cylinder.

The action can be reversed. Fluid can be pumped under pressure into the left end of the cylinder. The piston will be forced to the right.

Cylinders come in two basic designs, single- and double-acting. Double-acting cylinders can be either differential and non-differential cylinders. These three types of cylinders are shown in Figures 3-36, 3-37 and 3-38.

FLUID POWER SYSTEMS

The different units described in this chapter can be used in a variety of fluid power systems. These systems are useful especially where there is need for a great increase in force or speed. Hydraulics can be used to operate giant industrial presses, lathes, and small hand jacks for autos. Pneumatics can be used to operate stamping machines. They can operate automatic doors, air hammers, and impact wrenches. Fluid power provides the flexibility and mechanical advantages that make automation possible.

Chapter 4 ELECTRICAL POWER

BASIC PRINCIPLES

Scientists know that atoms are made up of very small particles. The **nucleus** or center of most atoms is made up of a group of **protons** and **neutrons.** Other tiny particles around the nucleus are called **electrons.** The particles of an atom are bound together by electrical energy. The electrons, protons, and neutrons are so small in size that the atom is made up mostly of space. Suppose the atom could be increased to the size of an average gymnasium. The nucleus would then be about the size of a marble located in the center. The electrons would be smaller particles whirling around at various distances from the nucleus. See Fig. 4-1.

Electrical Charge

Atoms in their normal state are made up of the same number of protons and electrons. When the number of electrons is not the same as the number of protrons, the atom has an electrical charge. Electrons are negative particles. Protrons are positive particles. If an atom has extra electrons, then it is **charged negatively.** If it is short of electrons, it is **charged positively.** Neutrons have no electrical charge. Therefore, they are neutral.

SOURCES OF ELECTRICAL ENERGY

Electrons flowing along a conductor are called **current electricity.** The electric current produced by batteries is called **direct current** (DC). This means that the electrons in wires or other conductors move in the same direction all the time the current is flowing. See Fig. 4-2. Most of the electricity used in homes is **alternating current** (AC). This means that the direc-

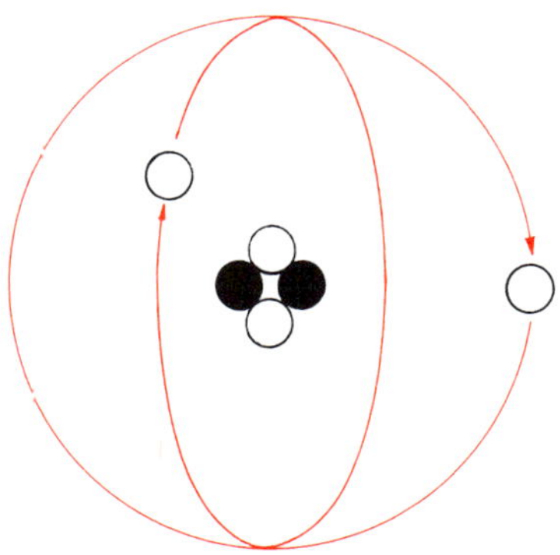

Fig. 4-1. In an atom, electrons move around the nucleus.

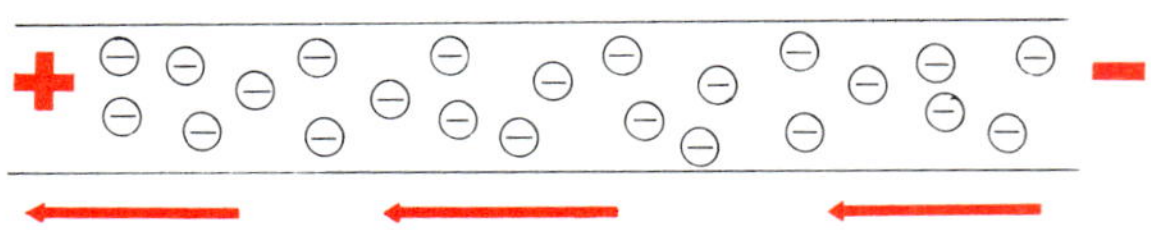

Fig. 4-2. Flow of electrons through wire.

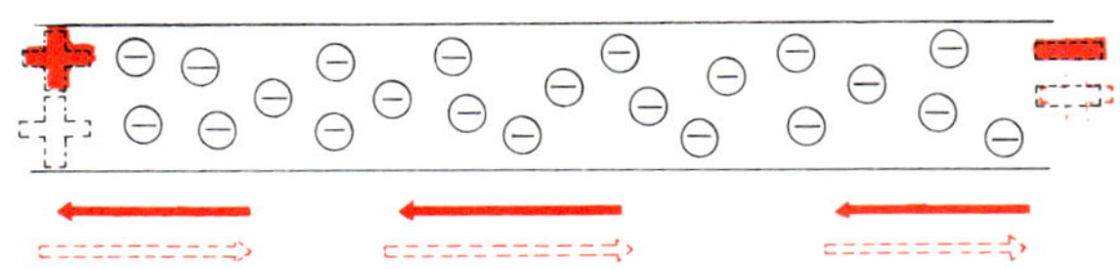

Fig. 4-3. Changing electron flow with alternating current.

tion in which the current flows changes several times a second. See Fig. 4-3. Electrons in the conductor move first in one direction and then in the other direction.

Alternating current is usually shown by a curved line. See Fig. 4-4. Notice that the current starts at 0. It reaches a peak in the positive direction and goes back to 0. It then reaches a peak in the negative direction and then moves back to 0. This movement of the current represents one **cycle.** Most of the alternating current in home use is a 60 cycle current. This means that the current changes direction 120 times per second while it goes through 60 cycles per second. This change in direction of alternating current is a very important factor in producing electric currents. These currents move through wires for hundreds of miles across the country. Also, the strength of alternating current can be changed.

ELECTRIC GENERATORS

If a wire moves through a magnetic field, the electrons in the wire are disturbed. This causes a current of electricity to flow through the wire. See Fig. 4-5. This is how electricity is produced for homes, factories, and cities. Most of the electricity in use is produced by generators. They are called **dynamos,** Fig. 4-6.

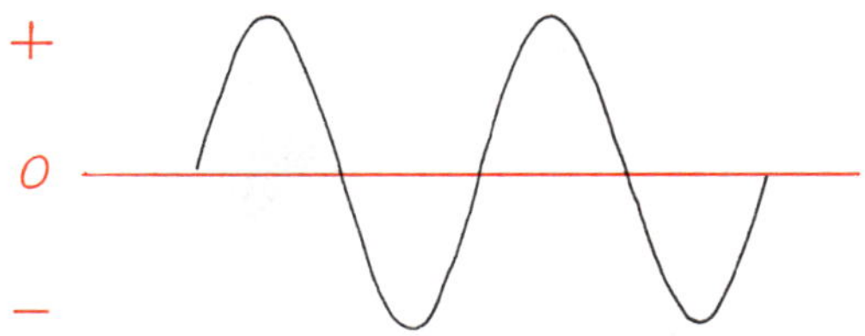

Fig. 4-4. Alternating Current

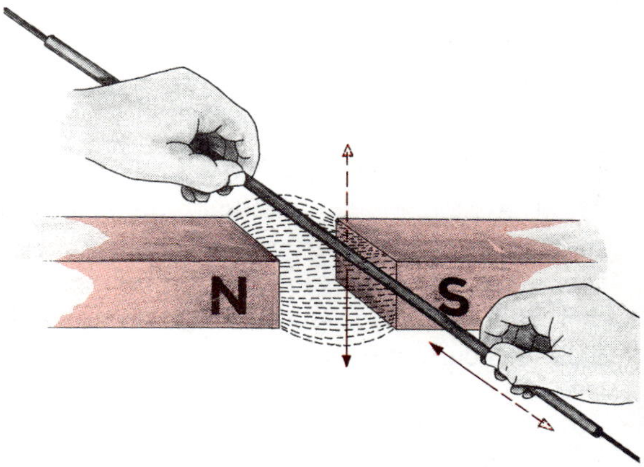

Fig. 4-5. Moving wire through a magnetic field.

Fig. 4-6. Electric Generator

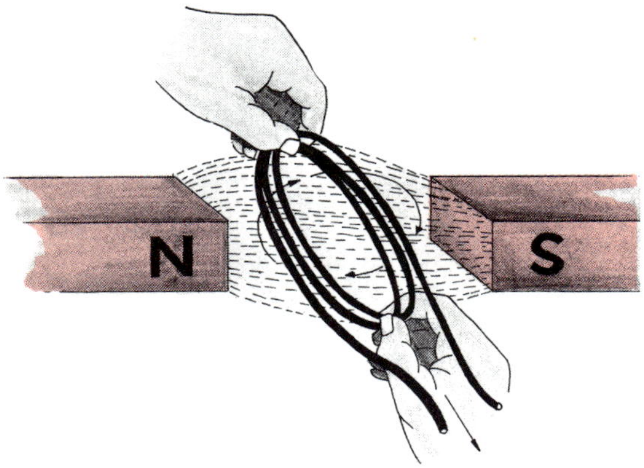

Fig. 4-7. Turning wire coil in magnetic field.

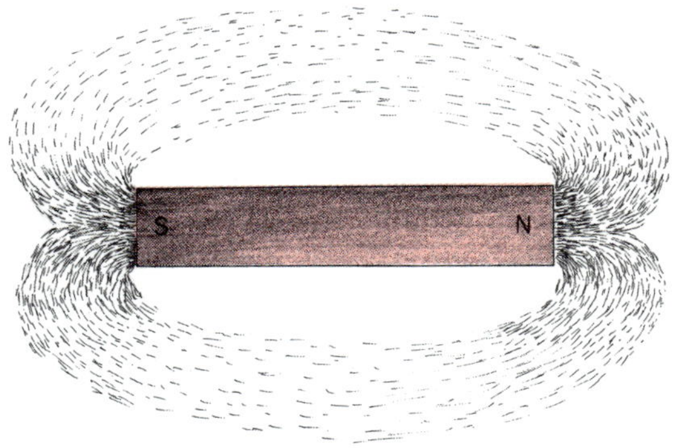

Fig. 4-8. Iron filings show magnetic field.

Increasing Current Strength

The amount of electricity produced by passing one wire through a magnetic field is very small. Suppose a long wire is rolled into a coil of wire and passed through a magnetic field. The current flow is then increased. The increase results from several wires cutting across the magnetic field. See Fig. 4-7. The current is also increased when the magnetic field is strengthened. This is because more electrons are put in motion. Also, the speed at which the wire coil is moved through the magnetic field will affect the voltage.

MAGNETISM

Electrical principles are based on magnetism. The first magnets known were natural stones that contained iron ore. These stones are called **lodestones.** The magnetic field surrounding a lodestone is rather weak. Natural magnets are not used commercially. Permanent magnets can be made that are much stronger. Special iron or steel is magnetized by using coils of wire and an electric current. Permanent magnets are used for telephone receivers and radio speakers. They are used for many other items, such as tack hammers and compasses.

Fig. 4-9. Poles of a Magnet

It is easy to see the field of force that surrounds a magnet. Hold a magnet under a sheet of stiff paper. Sprinkle iron filings on top of the paper and then tap the paper. The iron filings will line up with the forces set up by the magnetic field. See Fig. 4-8. You will note that the greatest fields of force are located at each end of the magnet. The ends of magnets are called **poles.** One pole is called the north pole (N), and the other is called the south pole (S), Fig. 4-9.

Behavior of Magnets

Suppose two bar magnets are placed on a smooth surface. The end of one is moved toward the end of the other. The magnets will either be **attracted** to each other or will be

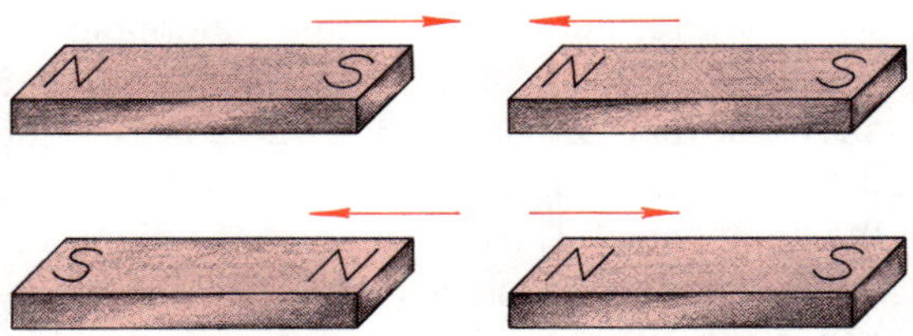

Fig. 4-10. Unlike poles attract; like poles repel.

forced apart (repelled) by each other. Their action will depend upon which ends are brought together, Fig. 4-10.

Electromagnets

Other types of magnets are even more useful than the permanent magnets. These are made by wrapping a coil of insulated wire around a iron core. When an electric current is sent through the coil of wire, a magnetic field of force is set up around the coil of wire. It is concentrated in the iron core. This type of magnet is called an **electromagnet.** See Figs. 4-26 and 4-27. Like the permanent magnet, it has a north and a south pole.

INDUCTION

Electricity can be transported for great distances. This is made possible by a principle of electricity called electrical **induction.** You have learned that when a current of electricity is passed through a coil of wire, a magnetic field of force is set up around the wires of the coil. As the current flows, the magnetic field moves out from the wires. When the current is stopped, the magnetic field moves back toward the wire and also stops. As an alternating current moves through the wire, it changes direction 120 times per second. This change in direction also causes the magnetic field to move back and forth. See Fig. 4-11. If a second coil of wire is placed near the first coil, an electric current will also flow through it. This occurs even though the two coils do not touch. The magnetic field moving back and forth from the first coil cuts across the wires of the second coil, Fig. 4-12. When a wire moves through a magnetic field, or when a magnetic field cuts across a wire, a current of electricity is produced in the wire. A current of electricity transferred from one coil to another is called **electrical induction.**

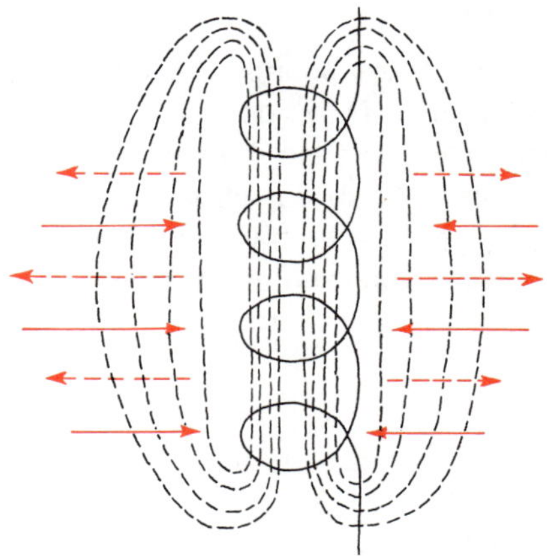
Fig. 4-11. Magnetic Field Movement

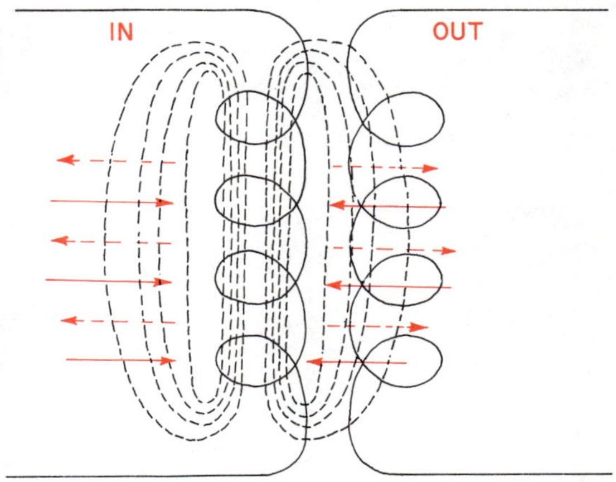

Fig. 4-12. Electrical Induction

Transformers

Transformers change the voltage strength of electricity. They do this by **induction.** Transformers increase the voltage of the alternating current carried across the country to cities, Fig. 4-13. A high voltage, however, must be reduced when it enters homes, Fig. 4-14.

Southwestern Gas and Electric Co.

Fig. 4-13. Electrical Transformers

Fig. 4-14. Transformer reducing high voltage.

Transformers can either increase or decrease voltage of a current. Figure 4-15 shows a transformer symbol. The coil carrying the electricity coming into the transformer is the **primary coil.** The coil in which the current is **induced** is the secondary coil. The secondary coil in the diagram has more turns of wire than the primary coil. This means that current going out of the transformer will be increased. This is a **step-up** transformer. The high voltage can travel over power lines to cities.

When the high voltage gets to a city, it goes through a **step-down** transformer, Fig. 4-16. The high voltage goes through the primary coil and **induces** current into the secondary coil. This steps-down the voltage so that the electricity can be used in homes.

CHEMICAL ELECTRICITY BATTERIES

Electricity can also be produced by chemical action. The dry cell and acid storage

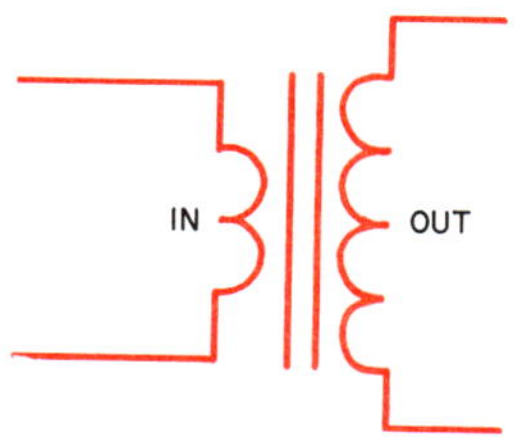

Fig. 4-15. Step-up transformer reduces voltage when it induces current from the primary to the secondary.

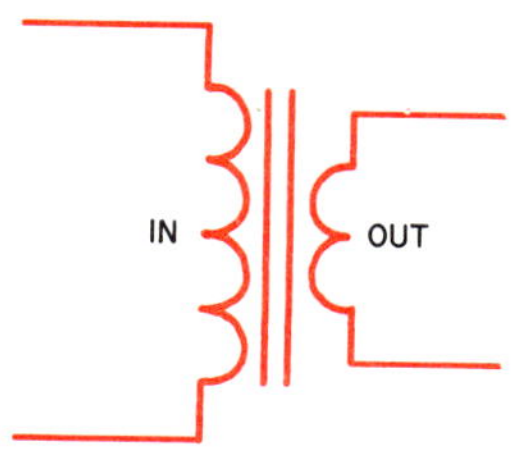

Fig. 4-16. Step-down transformer increases voltage.

batteries use chemical action to produce electricity. The dry cell batteries are known as **primary** cells. The lead-acid storage batteries are called **secondary** cells.

Primary Cell. The primary cell is often called a "dry cell battery." This leads to two misunderstandings. First, this form of electrochemical energy is **not dry.** It uses an ammonium chloride paste for the electrolite. It is moist rather than dry. The cell has a zinc shell or container. This shell serves as one of the electrodes. The cell also has a carbon rod which extends through the center of the cell. It serves as the other electrode. The top of the cell is sealed. This prevents evaporation of the moisture in the electrolite. Second, the primary cell is not made up of a group of cells, as the word "battery" implies. It has only one cell. The primary cell cannot be recharged.

In recent years, the primary cell has been improved. Alkaline electrolite and a mercury-oxide electrode are used rather than the zinc electrode and ammonium chloride electrolite. There is also the alkaline cell that has a negative electrode of manganese dioxide and the positive electrode of zinc. Both of these cells have certain features that make them improvements over the earlier dry cell. See Fig. 4-17.

Secondary Cells. Another type of electrochemical cell is the secondary cell. This type of cell can be recharged. It must first be charged by passing a direct current through it. This charging process brings about a chemical reaction that can later be reversed to yield a flow of electrons.

The secondary cell is often called a **storage cell.** This is not because the cell stores up electricity. It stores electrons that are released for use by the chemical reaction. The secondary cell most used now is the lead-acid cell. It has lead plates in a solution of sulphuric acid and water. See Fig. 4-18.

Union Carbide

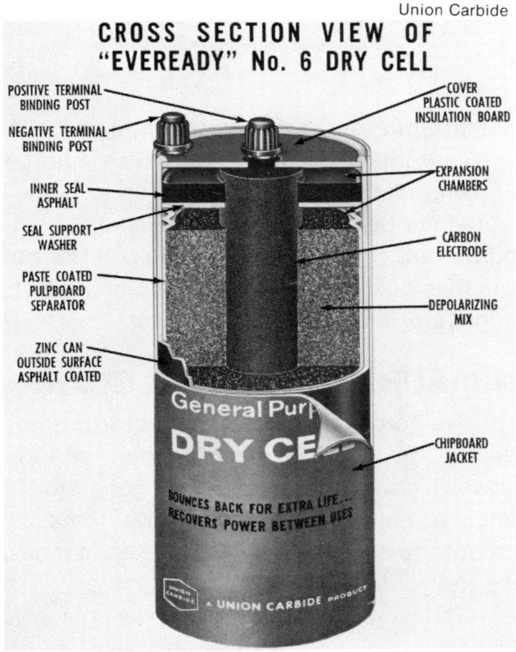

Fig. 4-17. Dry Cell Construction

The Electric Storage Battery Co.

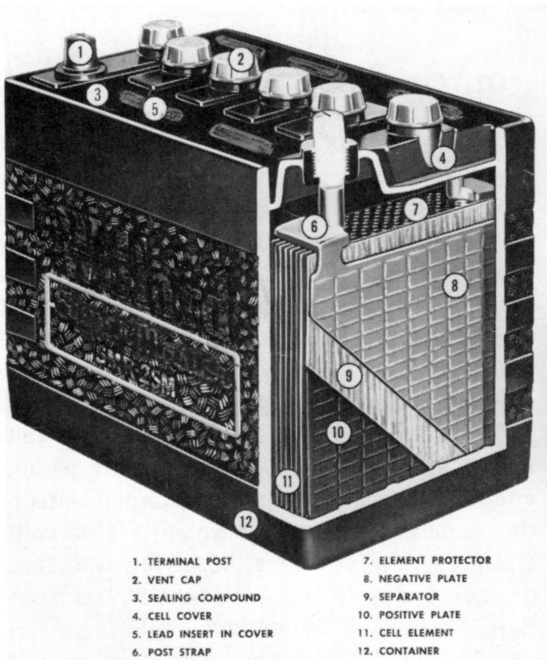

Fig. 4-18. Automotive Battery

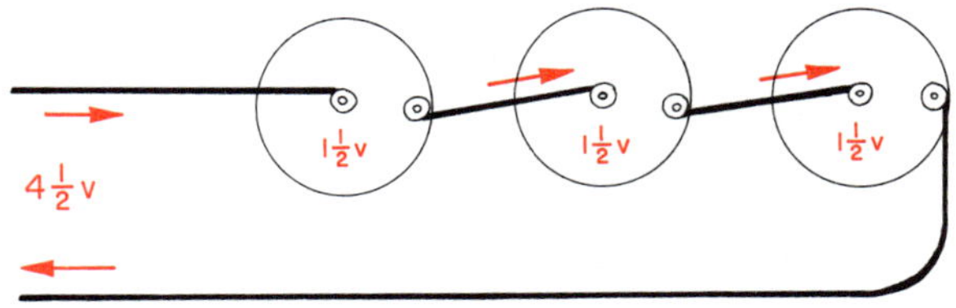

Fig. 4-19. Series Circuit Connection
Add the voltage of each cell when connected in series (4½ V here) for the circuit voltage. The amperage, or current flowing, stays the same for all parts of a series circuit.

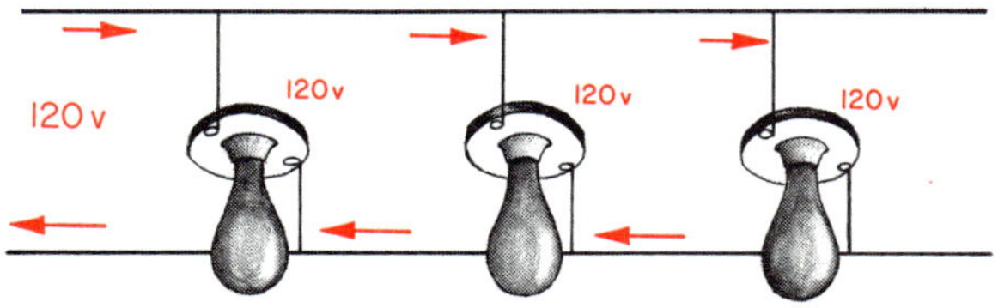

Fig. 4-21. Lights in a parallel circuit.

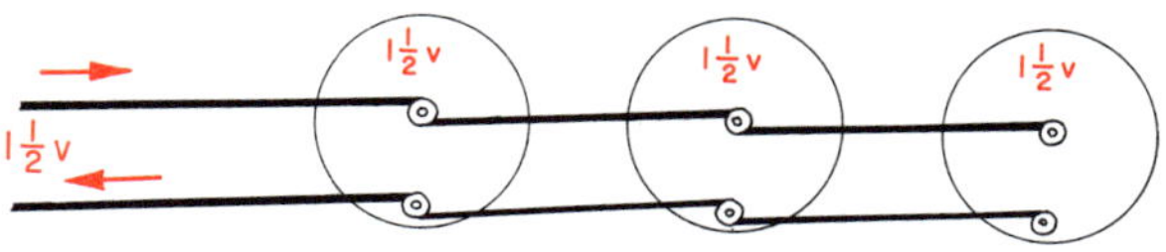

Fig. 4-20. Parallel Circuit Connection
The voltages stay the same, and the circuit amperage becomes .03A. The total circuit's current is the sum of the currents from each source.

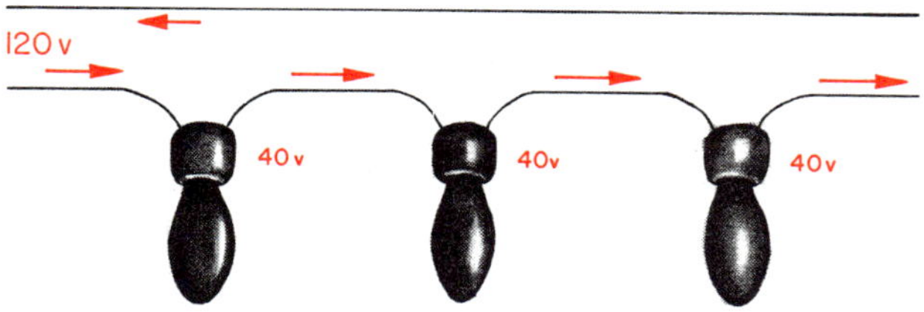

Fig. 4-22. Lights in a series circuit.

CIRCUITS

An electrical circuit is the path through which an electric current passes. There are two types of circuits. The dry cells shown in Fig. 4-19 are connected in a **series circuit.** Notice that the positive terminal of one cell is linked with the negative terminal of another cell. Each cell has about 1-1/2 volts. When each cell is connected in series as shown, the total voltage is 4-1/2 volts. The **amperage** (A), or amount of current flowing, stays the same. The dry cells shown in Fig. 4-20 are connected in a **parallel circuit.** The total voltage of the circuit remains the same as one dry cell which is 1-1/2 volts. The amperage becomes .03A. However, three dry cells used in this manner will last about three times as long as one dry cell.

The lights in homes are connected in parallel circuits. See Fig. 4-21. If one of the bulbs burns out, the others continue to burn because the electrical circuit has not been broken. Lights in a series circuit, Fig 4-22, would not be useful for homes. When one bulb burns out, the circuit is broken. All the bulbs will then go out. Some strings of Christmas tree lights are wired in a series circuit.

MEASURING ELECTRIC CURRENT

The force that causes an electric current to flow along a wire is supplied by a source of power. This may be a battery or generator. The force is called the **electromotive force.** It is referred to as **voltage.** The unit used in measuring the voltage is a **volt.**

The amount of current flowing is called **amperage.** The unit of measure is referred to as an **ampere** (amp).

The opposition to an electric current is called **resistance.** The unit of measure is an **ohm.** Some materials (such as copper and silver) let an electric current flow with little resistance. Such materials are called **electrical conductors.** Other materials (such as glass and rubber) offer a strong resistance to the flow of an electric current. They are called **insulators.**

Ohm's Law

In 1820, it was discovered that resistance, voltage, and amperage are related. This relationship is now known as Ohm's Law. Either the voltage, resistance, or amperage may be determined by mathematical formulas when the other two are known. In the following formulas, **E** means the voltage, **I** means amperage, and **R** - means resistance.

To find the voltage, multiply the amperage by the resistance.

$$E = I \times R$$

To find the amperage, divide the voltage by the resistance.

$$I = \frac{E}{R}$$

To find the resistance, divide the voltage by the amperage.

$$R = \frac{E}{I}$$

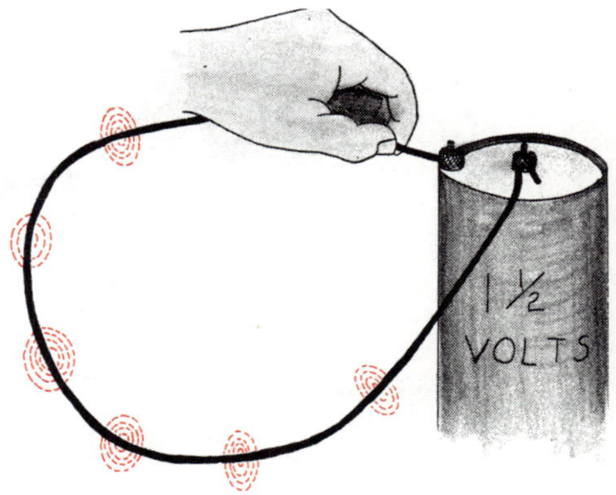

Fig. 4-23. Magnetic field formed by current carrying wire.

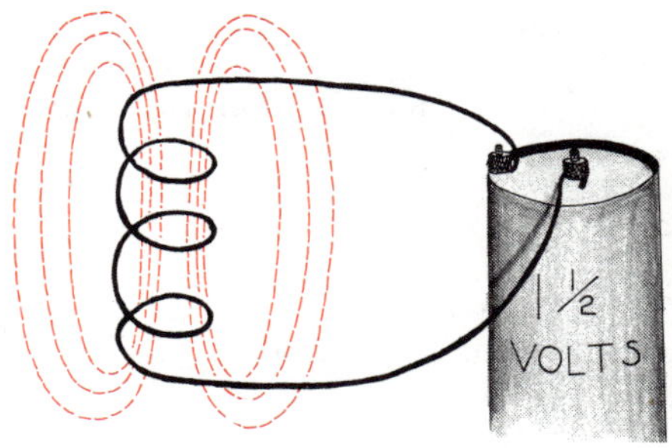

Fig. 4-24. Magnetic field formed by wire coil.

LEARNING ACTIVITIES

A wire passed through a magnetic field will disturb the electrons in the wire. It causes them to move and produce an electric current. The reverse of this is also true. When a current of electricity is passed through a wire, a magnetic field is set up around the wire. See Fig. 4-23. Because of this, hundreds of electrical devices can operate. Suppose a long wire is rolled in a coil and a current is passed through the coil. The strength of the magnetic force will be greatly increased. This is because the magnetic fields of all the loops are combined. See Fig. 4-24. If an iron core is placed through the coil of wire, the combined forces of all the wires center in the core. This changes the iron core to a magnet, Fig. 4-25. This type of magnet is called an **electromagnet.**

The electromagnet, like the permanent magnet, has a north and south pole. But the iron core of an electromagnet stays magnetized for only a very short time after the electric current is stopped. Also, the poles of electromagnets are changed when the direction of the flow of the current is changed.

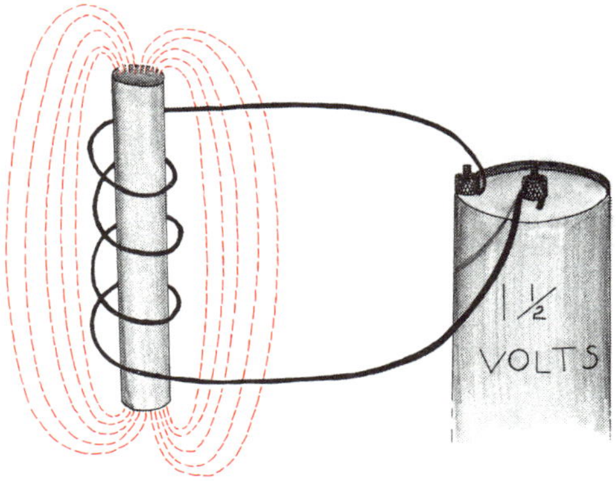

Fig. 4-25. Magnetic field concentrated in soft iron core.

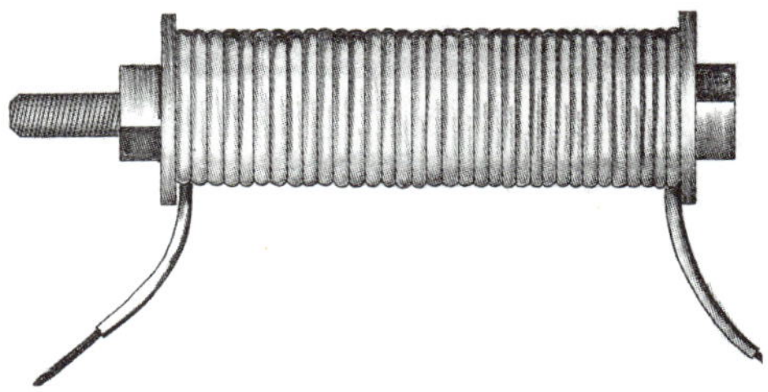

Fig. 4-26. Simple Electromagnet

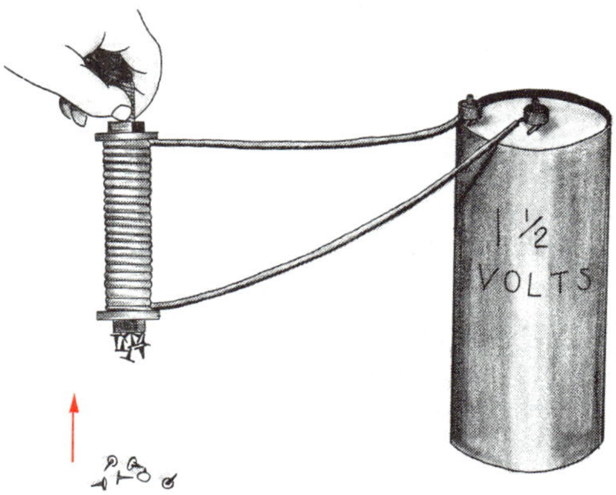

Fig. 4-27. Picking up tacks with an electromagnet.

You have learned that like poles of magnets repel each other and unlike poles attract each other. You can show this and other interesting facts by doing experiments with electromagnets.

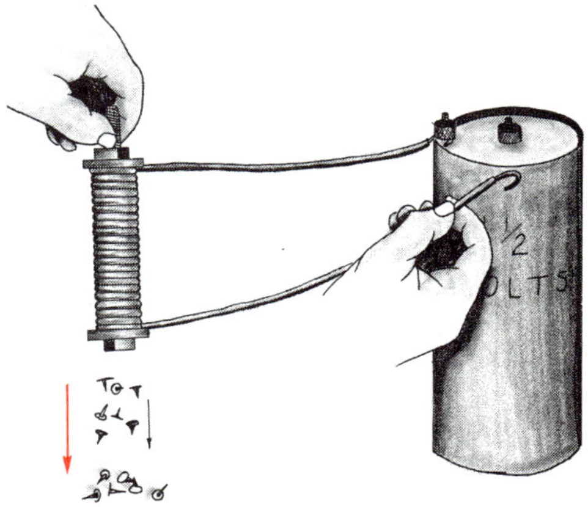

Fig. 4-28. Tacks fall when the current is cut off.

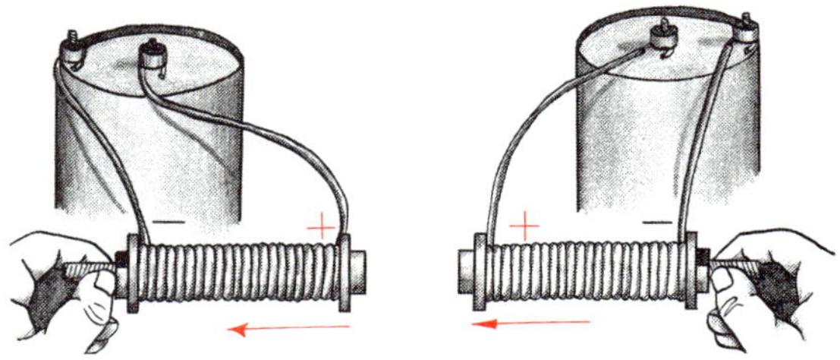

Fig. 4-29. Electromagnets connected to batteries.

Experiments

Experiment 1: Construct a simple electromagnet. Wrap about 20 turns of insulated wire around a bolt. See Fig. 4-26. Scrape about 1 in. (25 mm) of insulation from each end of the wire. Connect the bare ends of the wire to a common dry cell, Fig. 4-27. The electromagnet will attract and pick up small iron pieces, such as carpet tacks or wire brads. When one of the wires is removed from the dry cell, the electric current will no longer flow through the coil of wire. Tacks will drop from the end of the bolt, Fig. 4-28. This shows that the magnetic field of force is present only when an electric current is flowing through the wire.

Experiment 2: Construct a second electromagnet in the same way as the first. Connect

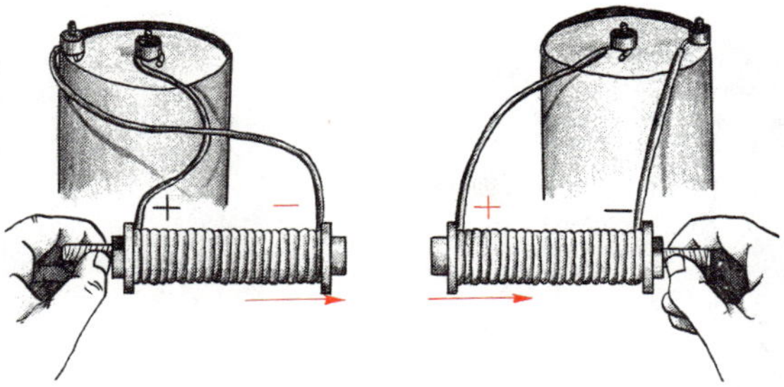

Fig. 4-30. Changing current direction in one magnet.

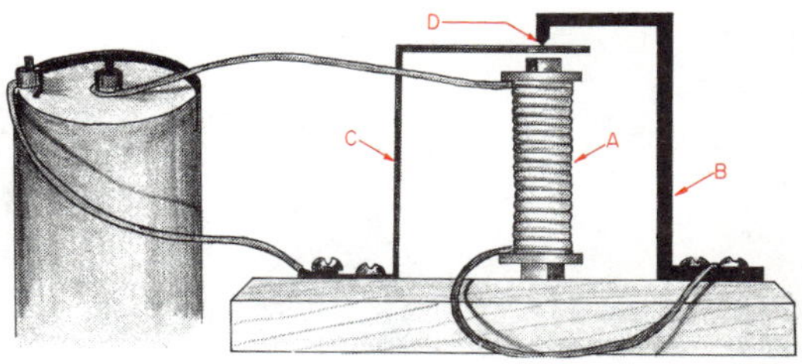

Fig. 4-31. Simple electric buzzer.

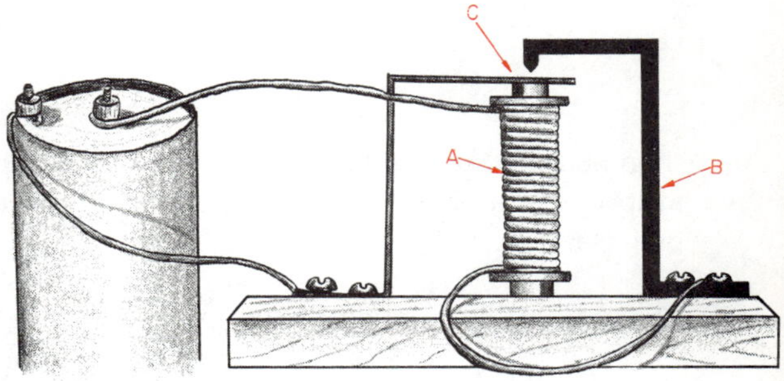

Fig. 4-32. Magnetic field attracts spring steel.

both magnets to dry cells. See Fig. 4-29. Bring one pole of each of the magnets toward each other. If the two poles repel each other, you know that they are alike. This is because like poles of magnets repel each other. If the two magnets attract each other, the two poles are opposite. This is because unlike poles of magnets attract each other. Turn one of the magnets around so that the opposite pole is used. The action of the two magnets is then reversed.

Experiment 3: Disconnect and switch the two wires on one of the dry cells, Fig. 4-30. You will see that changing the direction of the flow of the electric current also changes the poles of the electromagnets.

Electric Buzzer

The simple electric buzzer, Fig. 4-31, shows one use of electromagnets. Notice that one wire of the electromagnet **A** is connected to the dry cell. The other end is attached to the strip of metal **B.** The piece of spring steel **C** is attached to the other terminal of the dry cell. An electrical contact is made between the two strips of metal at **D.** When the flow of electricity is followed through the circuit, the current flows from the battery through the electromagnet **A.** Then it flows to the strip of metal **B.** The metal makes contact with the piece of spring steel **C.** The current then flows back to the battery to complete the circuit.

As the current flows through the coil of wire in electromagnet **A,** a magnetic field of force is set up in the iron core. The magnetic field of force attracts the strip of spring steel **C** and pulls it down. See Fig. 4-32. When the strip of metal **C** is pulled down, the electrical contact is broken between the two pieces of metal at point **D.** The current cannot flow because the circuit is broken. When the electric current stops flowing, the magnetic field of force centered in the magnet's iron core also stops. This allows the strip of metal **C** to spring back up. It again makes contact at point **D.** This starts the current flowing through the circuit to repeat the process. The strip of metal **C** vibrates rapidly back and forth between the electromagnet and contact point **D.** This produces a buzzing sound.

The buzzer can easily be changed to an electric bell. A small piece of metal is secured to the end of strip **C** to make a clapper. A bell is

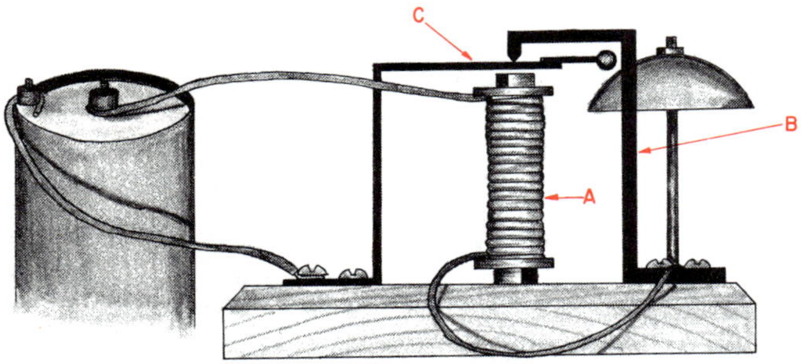

Fig. 4-33. Converting buzzer into a bell.

then attached to the wood base. Each time the strip of metal **C** vibrates, the clapper hits the bell. See Fig. 4-33.

Electric Motor

The operation of an electric motor is more complex than the operation of a buzzer or telegraph. But, if you now understand the basic principles of electricity, you can understand the principle by which a simple electric motor operates. An electric motor is very similar to an electric generator. However, it operates in reverse. The electric generator is turned by some source of power to produce electricity. The electric motor is turned by electricity to produce power for various uses. Figure 4-34 shows a very simple electric motor. The iron core is bent into the shape of a horseshoe. The field magnet **A** is made by wrapping a coil of wire around the core. The armature **B** is also an electromagnet. The ends of the coil of wire are attached to the commutator plates **C** and **D.** Both the armature and commutator plates are secured to a wood dowel. The dowel is suspended between two upright supports so that it will turn freely. The two brushes of the motor, **E** and **F,** are attached to the wood base. They rest against the turning commutator plates.

Fig. 4-34. Parts of an electric motor.

The electrical circuit going through the motor passes from the dry cell through the field coil to brush **F.** It goes through one plate of the commutator plate to brush **E.** Then it goes back to the dry cell to complete the circuit. As the current from the dry cell passes through the field coil **A,** a magnetic field is set up. It is centered in the iron core. One end of the horseshoe magnet becomes the north pole and the other the south pole. These poles remain the same at all times. This is because the direction of the current stays the same.

The poles of the armature electromagnet, however, change as the armature turns. The commutator plates turn as the armature turns. They are attached to the same wood dowel. As the plates turn, they switch the brushes with which they are in contact on each half turn. Every time the commutator switches brushes, the direction of the electric current passing through the armature coil changes. The commutator plates are arranged on the dowel so that the armature magnetic poles are unlike the field magnet poles when the armature is resting

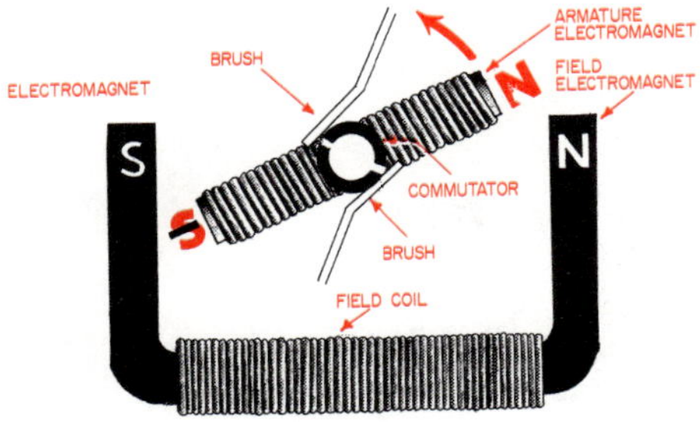

Fig. 4-35. Turning of an armature.

Fig. 4-36. Reversing current flow.

in the position shown in Fig. 4-35. These unlike poles attract each other. This causes the armature magnet to turn until the poles are in line.

The armature would stop at this point except for one reason. Just before the poles of the armature magnet come in line with the field magnet poles, the commutator plates have turned so that they switch brushes with which they are in contact. The change in brushes changes the direction in which the current is flowing through the armature coil. This changes the poles of the armature electromagnet. The poles of the field which are next to each other are now alike. These poles repel each other. This causes the armature to be pushed on around by the repulsion of like poles and the attraction of unlike poles. The commutator plates are placed on the dowel so that they change brushes at exactly the right time to keep the poles of the armature magnet turning.

SMALL GAS ENGINE

Chapter 5

A power system is made up of the complete **cycle of power** from source to use. This cycle includes three parts. They may be widely separated or brought together into a single unit. These parts are:

1. **Conversion of energy to power** — the source. Power sources include heat engines, hydroelectric power, nuclear power, batteries, wind, etc.
2. **Control and transmission.** Mechanics, fluids, and electricity are used to deliver power in the required form.
3. **Use** — changing heat into work. The power is used either as motion, heat, or light for the benefit of people. In the process, all of the energy put into the system is lost as heat.

Power is lost as heat energy. This is because all motion produces friction, and friction produces heat. Heat taken from the available power is lost from the power system. Energy cannot be destroyed. As a result, the amount of energy put into the system must equal the amount used, plus any loss at the source or during transmission. The forms of use (motion, heat, and light) all involve motion. They produce heat through friction. As a result, all the energy put into the system is lost in the process, or during use, as heat energy. Figure 5-1 shows this loss.

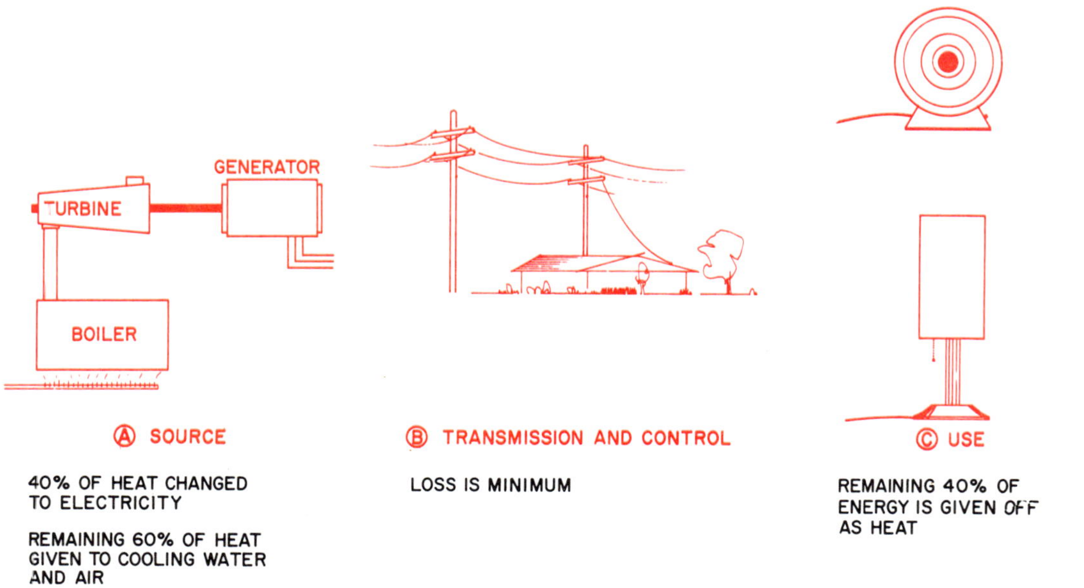

Fig. 5-1. Burning fuel releases energy to produce steam for the turbine. All this energy is lost as heat energy at the source or during transmission and use.

Power systems can cover great distances, such as the generation and use of electricity, Fig. 5-1. They can be quite compact. If a saw blade were attached to the small engine of Fig. 5-2, the system would be complete. The device could also be carried and used wherever needed.

To help you understand the operation of power systems, the operation of one system will be considered in detail. This system will consist of a small engine as the power source. A number of uses of this power source will be explained.

OPERATION OF THE SMALL ENGINE

Most small engines are produced as either two- or four-stroke cycle engines. The one shown in Fig. 5-2 is a two-stroke cycle engine. It can produce one horsepower.[1] The engine is quite small, measuring less than 8 inches (203 millimeters) in any direction. However, its operating systems are the same as large automobile engines. You will understand the operation of the automobile engine or other heat engines better after you study the small engine.

The small engine changes the chemical energy of fuel into motion. This change comes about because fuel is burned in a confined space. See Fig. 5-3. The heat produced by the burning fuel raises the temperature of the gas. The temperature increase raises the pressure. The pressure forces the piston down. This linear motion is changed to rotary motion and put to work.

A number of **subsystems** are needed to operate the engine. These subsystems work together to bring the proper mixture of air and fuel into the cylinder.

Small gasoline engines can do many different types of work. They provide power for lawnmowers, golf carts, boats, minibikes, water pumps, motorcycles, portable generators, power garden tools, chain saws, and snowmobiles.

SI Metric Conversion

1. In SI metrics, the measurement for power is watts (W). One horsepower equals 746 watts or 0.746 kilowatts.

Ohlsson & Rice, Inc.

Fig. 5-2. The small engine is a portable power source. It can be used to perform work as needed.

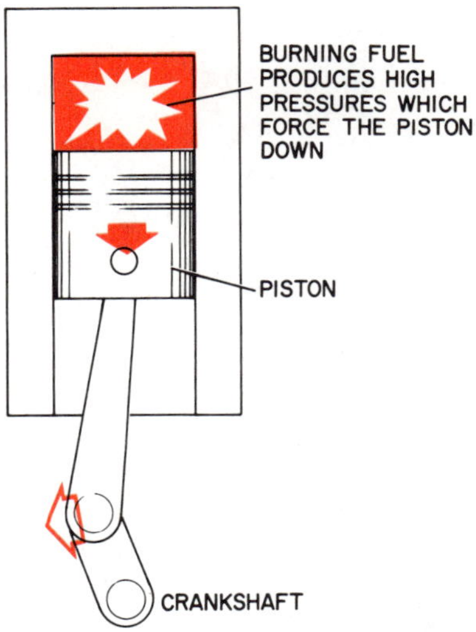

Fig. 5-3. Pressure from high temperature gases forces the piston down on a power stroke.

Small gasoline engines come in many sizes, shapes, and types. However, there are only two basic differences in all models. The engines are either **water-cooled** or **air-cooled.** They are either **two-stroke cycle** or **four-stroke cycle.**

The most common small engines are air-cooled. They are single-cylinder, two- and four-stroke cycle engines. They produce between one and 15 horsepower (746 and 11 200 watts).

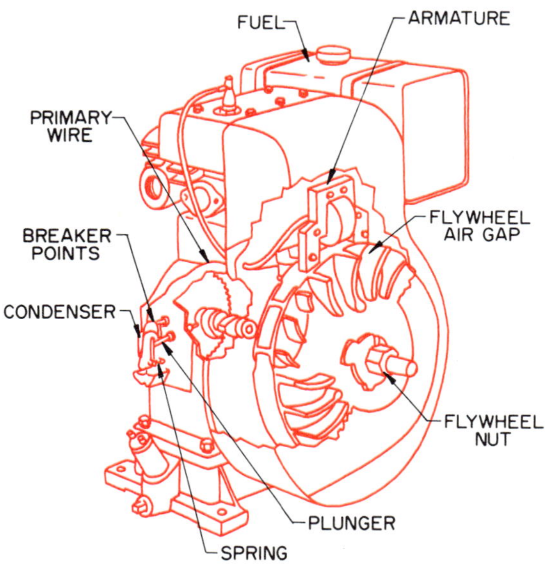

Fig. 5-4. Fuel, ignition, and lubrication system in a small gasoline engine.

The two- and four-stroke cycle gasoline engines are alike in many ways. They both have fuel, ignition, and lubrication systems. See Fig. 5-4. They also have a governor which limits the speed. Many of the parts are also similar. These will be discussed later in this chapter.

The two- and four-stroke cycle engines are reciprocating engines. This means that the motion of the piston is back and forth in a straight line. The back and forth motion is changed into rotary motion by the crankshaft, Fig. 5-3. Rotary motion is needed to operate the machines just mentioned.

A **stroke** is the motion of the piston in one direction. A new stroke is started each time the direction of the motion is changed.

The pressure for moving the piston is developed from the burning fuel mixture. The burning mixture causes an expansion of gases within the cylinder. An engine of this type is also called an **internal combustion engine.**

TWO-STROKE CYCLE ENGINES

A two-stroke cycle engine involves fuel intake, compression, power, and exhaust. However, the complete cycle takes place during one revolution of the crankshaft. The two-stroke cycle engine uses intake and exhaust ports. The crankcase is sealed.

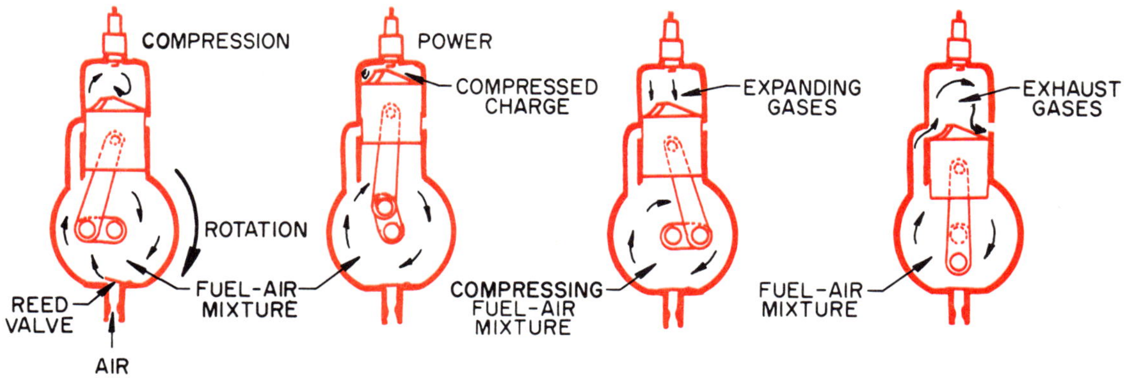

Fig. 5-5. One cycle in the two-stroke cycle engine requires one revolution of the crankshaft.

Up-Stroke of the Piston

As the piston moves up, two important things are occurring. The piston is compressing the fuel mixture in the cylinder. At the same time, it is creating a partial vacuum in the crankcase, Fig. 5-5A. Atmospheric pressure forces the reed valve to open. This allows a fuel mixture to enter the crankcase.

Down-Stroke of the Piston

As the piston reaches the top of the stroke, an electrical spark ignites the fuel mixture. The fuel mixture burns, expands, and pushes the piston down on the power stroke. See Fig. 5-5B. The piston moves down, putting pressure on the fuel mixture in the crankcase, Fig. 5-5C.

The piston uncovers the exhaust port first, letting the burned fuel mixture escape. Then the intake port is uncovered. This allows the compressed fuel mixture in the crankcase to flow into the cylinder. A deflector on top of the piston directs the incoming fuel mixture up and away from the exhaust gases. See Fig. 5-5D. The piston moves up. The new fuel mixture is trapped, and the cycle starts over again.

FOUR-STROKE CYCLE ENGINES

The four-stroke cycle engine has a cycle consisting of four strokes of the piston and two complete turns of the crankshaft. The four strokes are intake, compression, power, and exhaust.

Intake Stroke

During the intake stroke, the piston moves down, the exhaust valve is closed, and the intake valve is open. See Fig. 5-6A. The downward movement of the piston creates a partial vacuum. This allows atmospheric pressure to push a mixture of gasoline and air through the open intake valve. When the piston reaches the bottom of the stroke and starts up, the intake valve will close.

Compression Stroke

With the intake and exhaust valves closed, the fuel mixture is trapped in the cylinder. The piston moves up. The mixture is compressed into a space about 1/10 as large as it would ordinarily occupy. See Fig. 5-6B.

Power Stroke

As the piston reaches the top of the stroke, the compressed fuel mixture is ignited by an electrical spark. The burning fuel mixture expands rapidly and pushes the piston down on the power stroke. Both valves remain closed. See Fig. 5-6C.

Exhaust Stroke

Just before the piston reaches the bottom of the power stroke, the exhaust valve opens. The intake valve remains closed. The piston moves up, forcing the burned gases out the exhaust valve, Fig. 4-6D. As the piston reaches the top of the exhaust stroke, the intake valve opens and the exhaust valve closes.

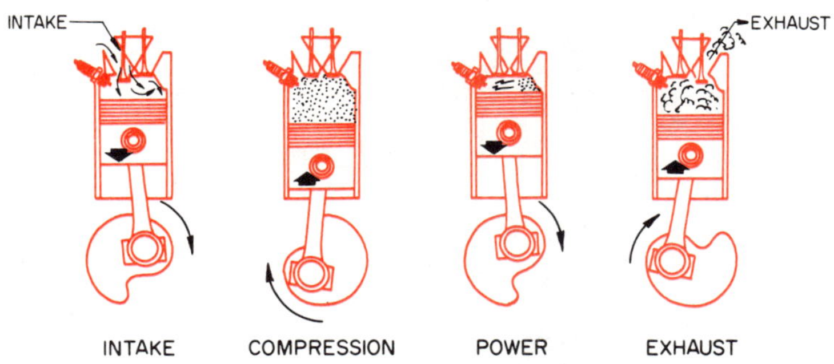

Fig. 5-6. One cycle in the four-stroke cycle internal combustion engine takes two revolutions of the crankshaft.

FUEL SYSTEMS

The fuel system must supply the engine with the right mixture of gasoline and air. The components of the fuel system have various functions. Gasoline is stored in the fuel tank. Fuel lines alone, or fuel lines and a fuel pump, deliver the gasoline to the carburetor. The gasoline is vaporized and mixed with air by the carburetor. The mixture is then supplied to the intake of the engine, Fig. 5-7.

FUEL DELIVERY SYSTEM

Four basic types of systems are used to deliver the gasoline to the carburetor. They are the (1) gravity feed system, (2) fuel pump or pressure system, (3) vacuum system, and (4) the pressurized fuel tank. The first two systems are the most common fuel delivery systems used on small gasoline engines. They are described below. The vacuum system is less commonly used now. The pressurized fuel tank is used by several outboard engine manufacturers.

Gravity Fuel System

Gravity is the force that carries the gasoline from the fuel tank to the carburetor. The fuel tank in this system must be mounted above the carburetor, Fig. 5-8.

Fuel (or Pressure) Pump System

Two major types of fuel pumps are used on small gasoline engines. These are the mechanical fuel pump and the vacuum fuel pump. No matter which of these systems is used, the fuel pump is most often mounted below the carburetor.

Mechanical Fuel Pump System

The mechanical fuel pump operates when the lever is moved up and down. See Fig. 5-9A. This motion comes from a cam in the engine. When the lever moves down at the cam end, the diaphragm moves up. This is shown in Fig. 5-9B.

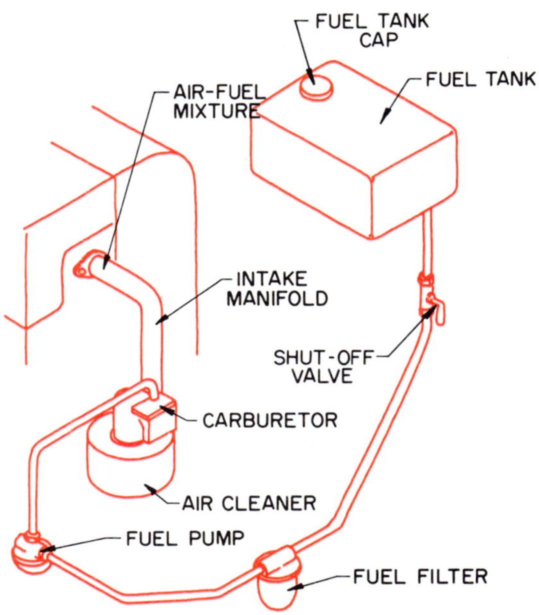

Fig. 5-7. Fuel supply system.

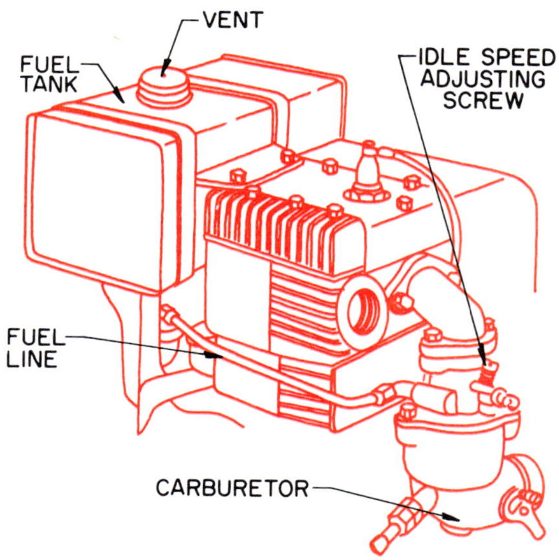

Fig. 5-8. Fuel tank is located above the carburetor.

The motion of the diaphragm forces fuel out to the carburetor. When the cam lifts the lever, the diaphragm moves down, causing a lower pressure in the pump. Atmospheric pressure on the fuel in the tank forces it into the fuel pump. The intake and outlet valves keep the fluid flowing in one direction. Since the fuel is pumped to the carburetor, the fuel tank is often mounted below the carburetor.

Vacuum Fuel Pump System

In four-stroke cycle engines, the vacuum fuel pump is operated by the intake manifold. In two-stroke cycle engines, it is operated by crankcase pressure. Only the crankcase system will be explained. See Fig. 5-10A. A diaphragm is placed across a chamber in the lower portion of the carburetor. The top half of this chamber is connected to the engine crankcase (dry chamber). The lower chamber (wet or pulsation chamber) is connected to the fuel tank by one line and to the carburetor by a second passage. The diaphragm contains two flapper valves. One controls the outlet port. The other controls the inlet port.

As the piston moves up, the partial vacuum in the crankcase pulls the diaphragm up. See Fig. 5-10B. This upward motion lets atmospheric pressure push gasoline through the inlet valve.

As the piston moves down, the pressure created in the crankcase pushes the diaphragm down, Fig. 5-10C. This closes the inlet valve and pushes the gasoline through the outlet valve to the carburetor.

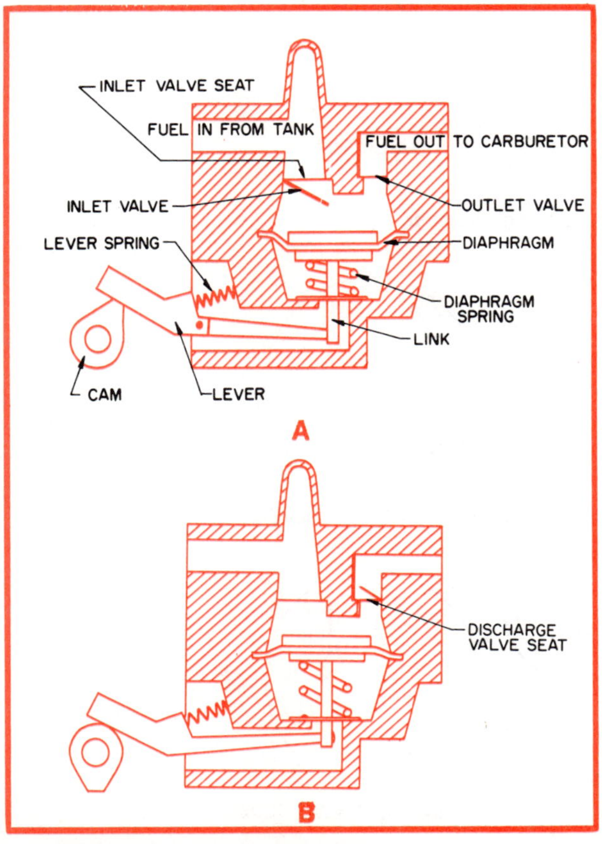

Fig. 5-9. Fuel pump operation.

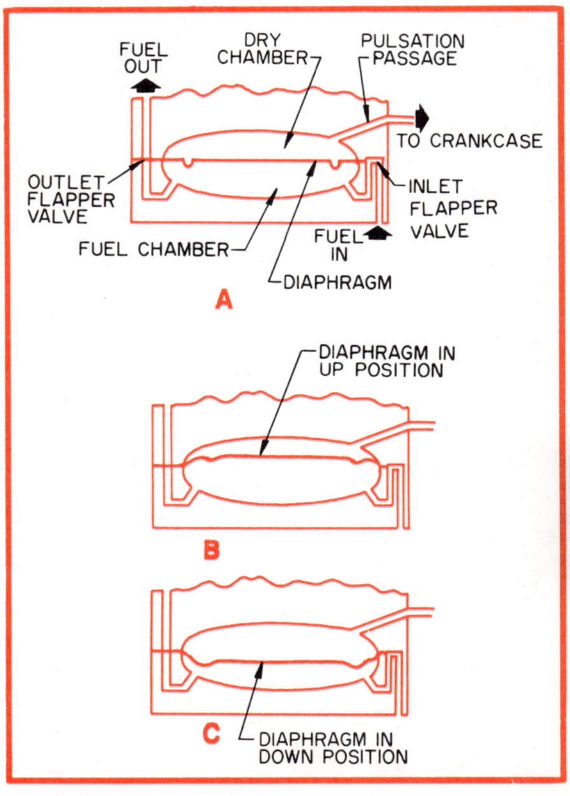

Fig. 5-10. Vacuum fuel pump.

In a four-stroke cycle engine, the diaphragm dry chamber is connected to the intake manifold, Fig. 5-11. The partial vacuum created on the intake stroke pulls on the diaphragm, covering the dry chamber. A spring pushes on the diaphragm between intake strokes.

CARBURETORS

Float-Type Carburetor

One of the most common carburetors used on small gasoline engines has a float that controls the fuel flow to the carburetor. As the fuel level drops in the float chamber, the float will move down, Fig. 5-12A. This allows the float needle to move down and open the inlet valve.

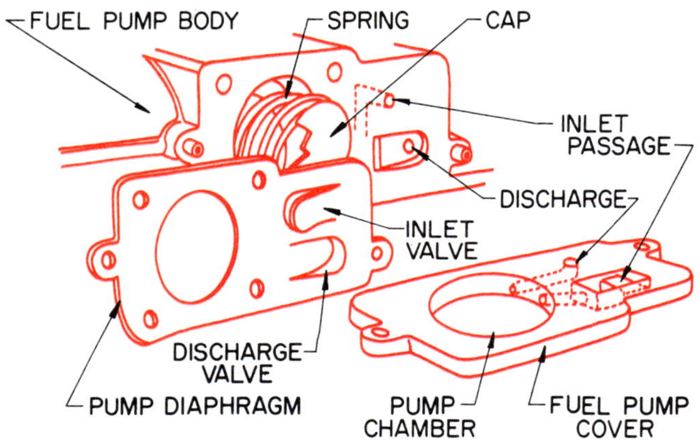

Fig. 5-11. Diaphragm covering chamber.

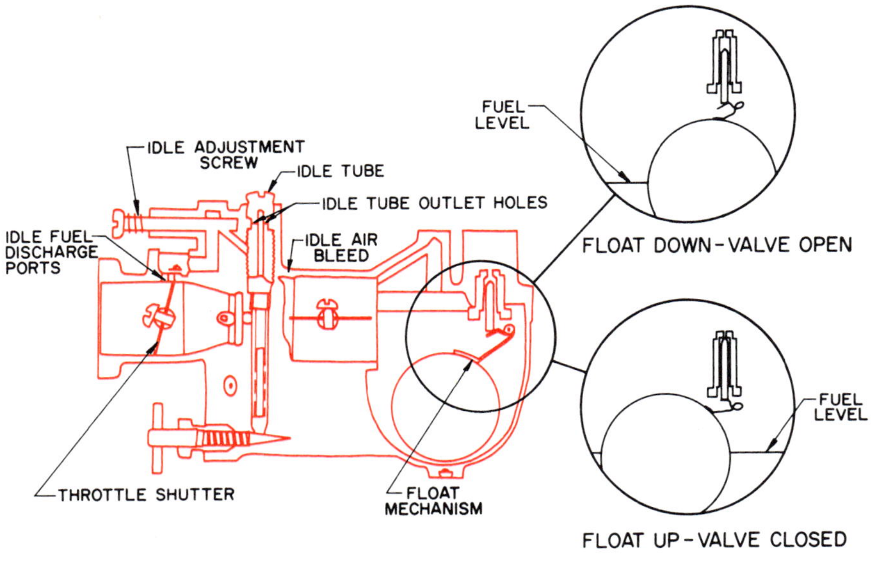

Fig. 5-12. Float carburetor system.

Diaphragm-Type Carburetor

The carburetor having a diaphragm can operate in any position. The float-type carburetor must stay level. A diaphragm is placed across a chamber, dividing it into a wet chamber and a dry chamber. See Fig. 5-13. The wet chamber is connected to the fuel tank or fuel pump. The dry chamber is vented to atmospheric pressure. The wet chamber contains a lever, spring, needle valve, and seat. As the fuel leaves the wet chamber, the diaphragm moves up. It lets the needle valve move down to open the inlet valve.

The fuel flows into the wet chamber and pushes the diaphragm down. The downward motion of the diaphragm raises the needle valve which closed the inlet and stops the flow of fuel.

VACUUM CARBURETORS

The fuel tank is mounted to the bottom of the vacuum carburetor, Fig. 5-14. A pickup tube sticks out of the carburetor into the tank. The partial vacuum that is created in the carburetor lets atmospheric pressure push fuel up the tube into the carburetor. The distance that the fuel can be lifted is limited. A shallow fuel tank must therefore be used.

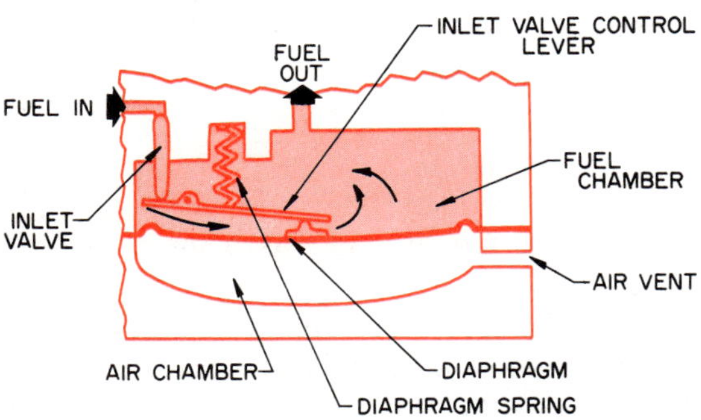

Fig. 5-13. Divided chamber.

Some vacuum-feed carburetors use a deep fuel tank and a vacuum-operated fuel pump. See Fig. 5-15. The gasoline is pumped from the tank into a fuel chamber having a constant level.

THEORY OF CARBURETION

The carburetor is a mixing valve that supplies the engine with a vaporized mixture of gasoline and air. This is done by creating a partial vacuum in the carburetor and forcing the gasoline through small openings or jets. The

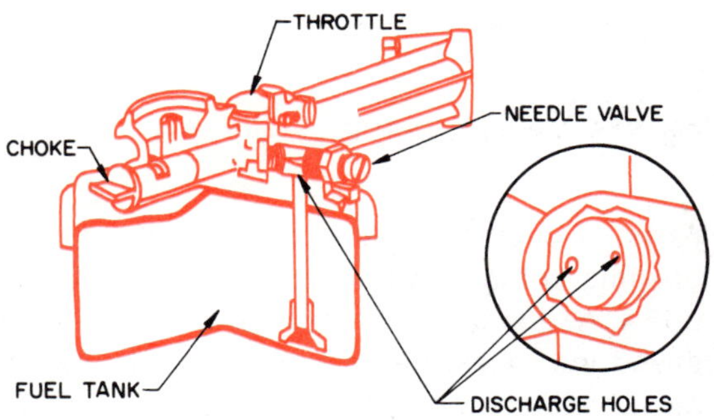

Fig. 5-14. Fuel tank is located below the carburetor.

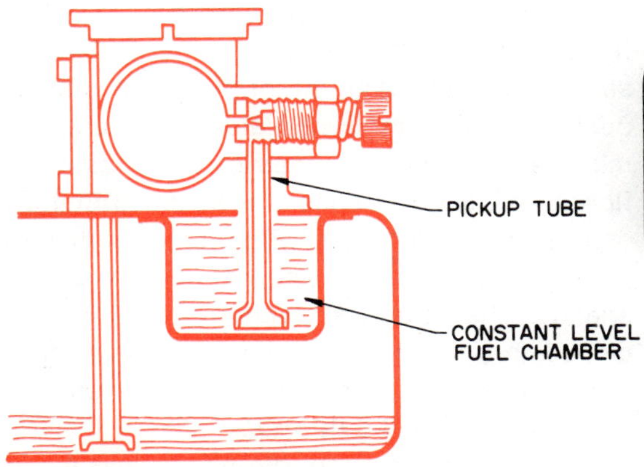

Fig. 5-15. Vacuum-operated pump.

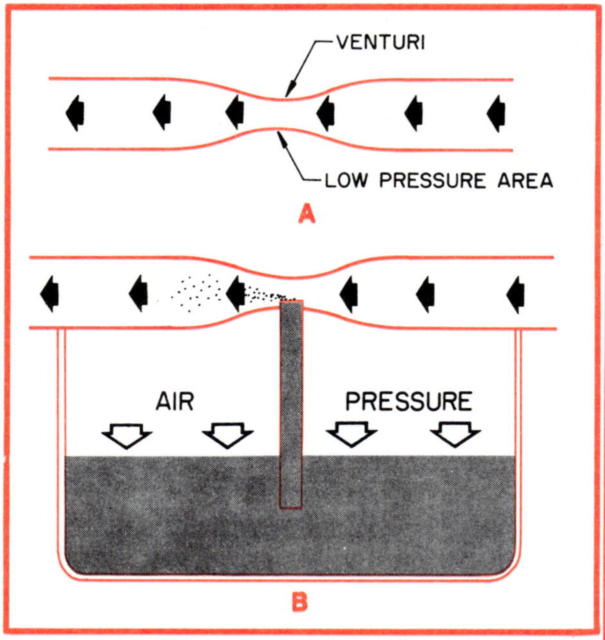

Fig. 5-16. Venturi principle.

partial vacuum is created by placing a narrow opening in the carburetor called a **venturi,** Fig. 5-16A. As the air flows through the venturi, its speed increases. Its pressure drops below atmospheric pressure. Atmospheric pressure on the gasoline will push the fuel up the tube and out the openings located in the low pressure areas at the venturi. See Fig. 5-16B. This is the main discharge tube. It will supply the engine with fuel at high speeds.

A **throttle plate** controls the amount of air flowing through the carburetor, Fig. 5-17A. The air controls the amount of fuel that will be pushed out the carburetor jets. The air and fuel mixture controls the speed of the engine. Air flow at idle and low speeds is so small that the engine will tend to stall. The idle circuit supplies the needed fuel and air mixture at low speeds. The strongest vacuum is behind the closed throttle at low speeds. Fuel is pushed out the idle jet. It mixes with the small amount of air going past the throttle.

Air is bled into the idle circuit and mixes with the fuel. The air-bleed system helps atomize the fuel. It also makes the carburetor easier to adjust.

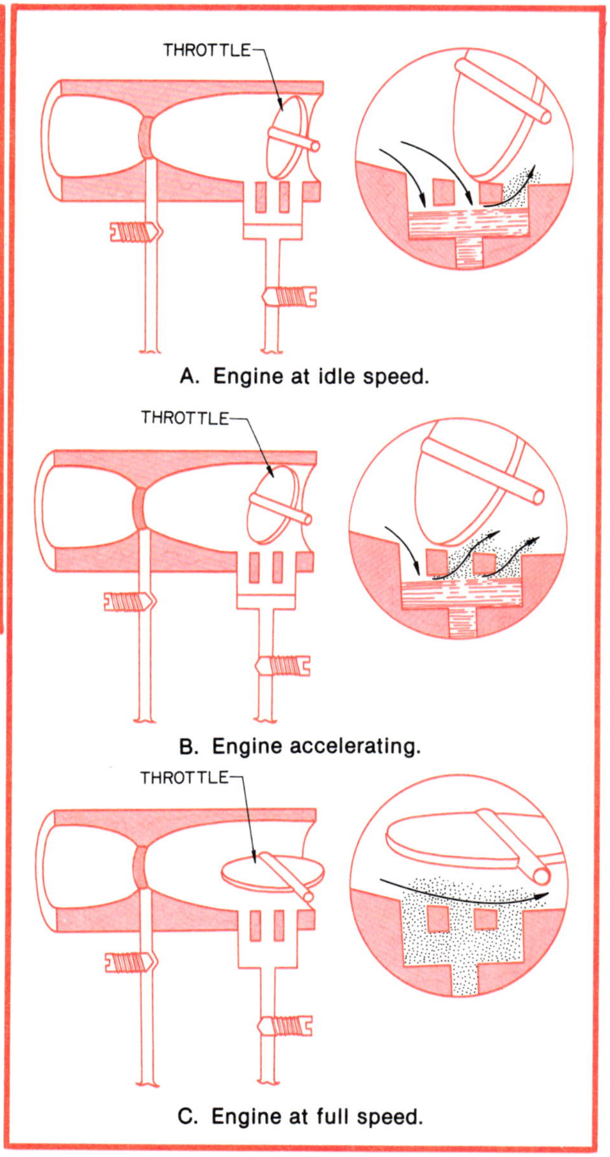

Fig. 5-17. Engine throttle operation.

As the throttle is opened, a second jet enters the area of strongest vacuum. See Fig. 5-17B. Fuel flows out both the idle and the secondary idle jets.

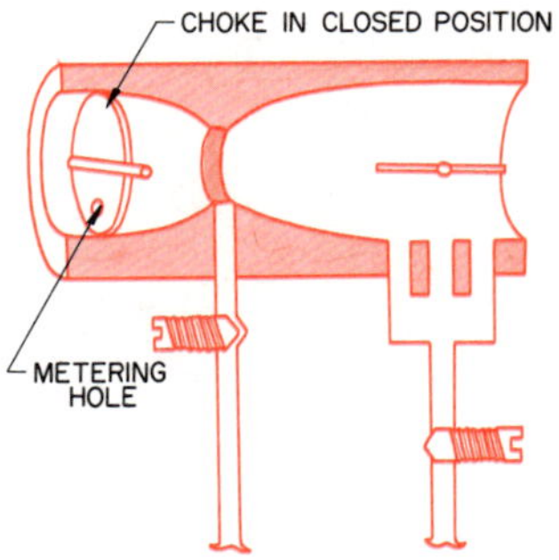

Fig. 5-18. Choke plate in carburetor.

Opening the throttle more causes the air bleed hole to have fuel pushed through it. This delivers the necessary fuel for moderate speeds. When the throttle is open full, the main jet will supply the engine with most of the needed fuel. See Fig. 5-17C.

A choke plate is located near the air intake of the carburetor, Fig. 5-18. When the choke is closed, very little air can enter the carburetor. A metering hole in the choke controls the amount of air that enters the carburetor. All the jets are in the area of strongest vacuum. Fuel is pushed out all the holes, supplying the engine with a large amount of fuel and a small amount of air. This is a rich fuel mixture. It is needed to start a cold engine.

AIR FILTERS

Air filters are placed on the intake of the carburetor. They remove unwanted particles from the air. There are three major types of air filters. They are (1) oil bath, (2) wet foam, and (3) dry type. See Fig. 5-19.

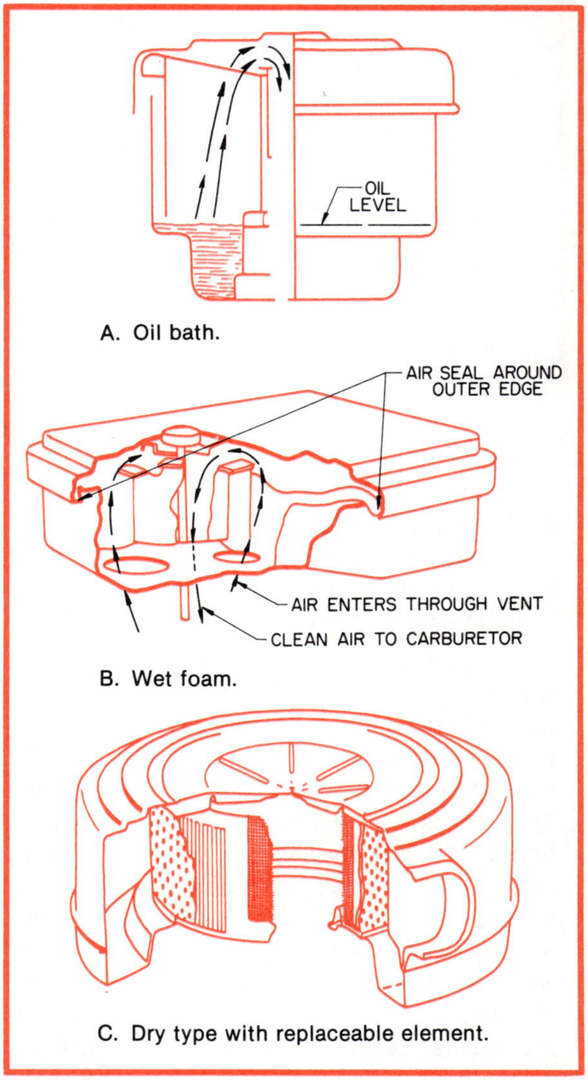

A. Oil bath.

B. Wet foam.

C. Dry type with replaceable element.

Fig. 5-19. Air filters.

IGNITION SYSTEM

The small gasoline engine uses a spark ignition system. The electrical system produces the high voltage needed to cause an arc across the air gap of the spark plug. The magnetic ignition system is the more common. This type of electrical system is commonly called a **magneto.**

Operation of a magneto is based on several rules of electricity:

1. A complete circuit is needed to have electric current.

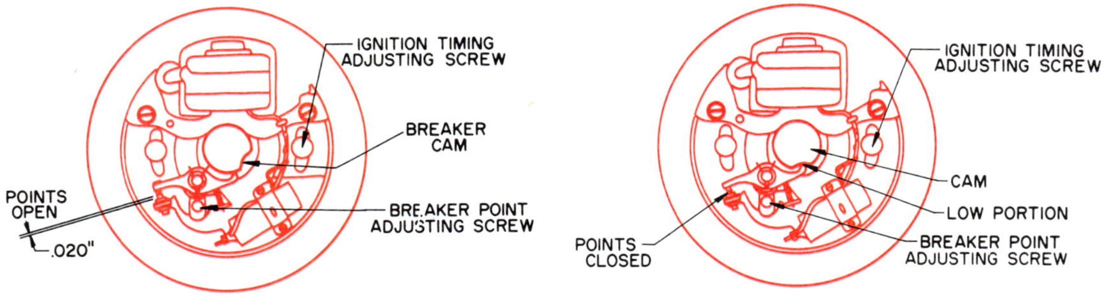

A. Magneto with points open. B. Magneto with points closed.

Fig. 5-20. Flywheel magneto.

U.S. Customary	SI Metric
.020 in.	0.5080 mm

2. Electricity is caused by moving a magnet past the coil of wire.
3. Electric current flow in a coil causes a magnetic field to build up around the coil.

A permanent magnet is mounted in the flywheel, Fig. 5-20A. This magnet is rotated past the ignition coil. A voltage is induced into the primary winding of the ignition coil as the magnet passes. The contact points are closed. This makes a complete circuit for current to flow in the primary section. The current flow through the primary coil causes a magnetic field to build up around it. The primary current will stop when the magnet is moved away. Or it will stop when the breaker cam causes the contact points to open. See Fig. 5-20B. Opening of the points causes a rapid collapse of the magnetic field. The collapsing lines of force affect the secondary of the ignition coil by inducing a large voltage. This voltage will cause an arc across the air gap of the spark plug. The arc or spark ignites the fuel and air mixture. This causes the piston to move, thereby rotating the flywheel and the breaker cam. The contact points close. The magnet moves past the ignition coil to begin another cycle.

A high voltage is also induced into the primary circuit. The capacitor stores this voltage, helping the magnetic field to collapse rapidly. The capacitor also prevents arcing across the ignition points. Arcing would damage the points.

LUBRICATION

Three basic methods are used to lubricate the many moving parts in a small gasoline engine. They are (1) splash, (2) pressure, and (3) mixing the oil with the gasoline. The last method is used with two-stroke cycle engines.

Splash Lubrication Method

Engines that use the splash lubricator system will use one of two methods. A **horizontal shaft engine** will have an oil dipper connected to the connecting rod. See Fig. 5-21. The dipper moves through the oil as the crankshaft rotates. It splashes the oil onto the engine parts. A **vertical shaft engine** will have an oil slinger that consists of a spear with blades. See Fig. 5-22. The spear is turned by the crankshaft gear. Its blades sling oil to the engine parts.

Pressure Lubrication Method

The pressure type of lubrication system uses an oil pump. The pump is operated by the crankshaft, Fig. 5-23. A plunger moves up, forcing oil through drilled passages to the bearings.

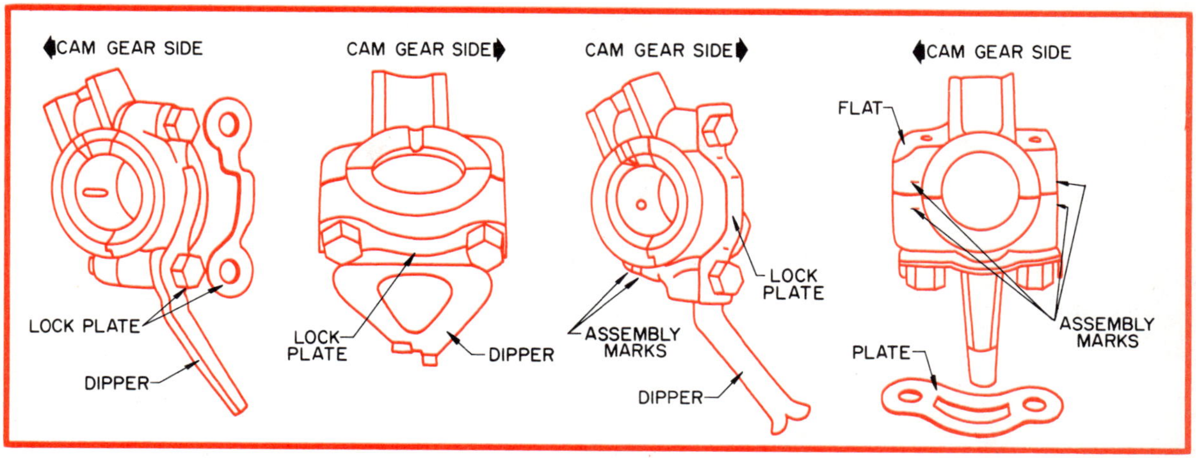

Fig. 5-21. Installation of a connecting rod on horizontal crankshaft engines.

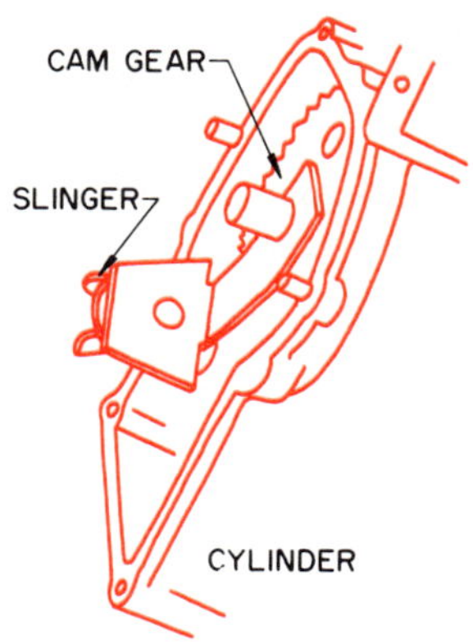

Fig. 5-22. Small engine oil slinger.

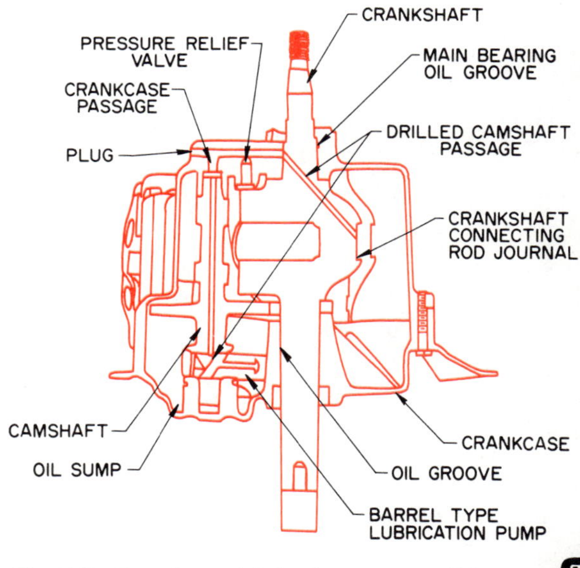

Fig. 5-23. Barrel-type lubrication system oil flow.

The other parts are lubricated by the oil that is slung around by the crankshaft and camshaft.

An engine should not be operated at top speed at all times. Most small gasoline engines are fitted with some type of **governor control.** This control can be set to maintain a certain speed range. The two types of governors are pneumatic and mechanical.

GOVERNORS

Pneumatic

The pneumatic or air vane governor is located near the flywheel, Fig. 5-24. As the engine speed increases, the amount of air flowing over the air vane is increased. The air vane pushes on the throttle linkage. This closes the

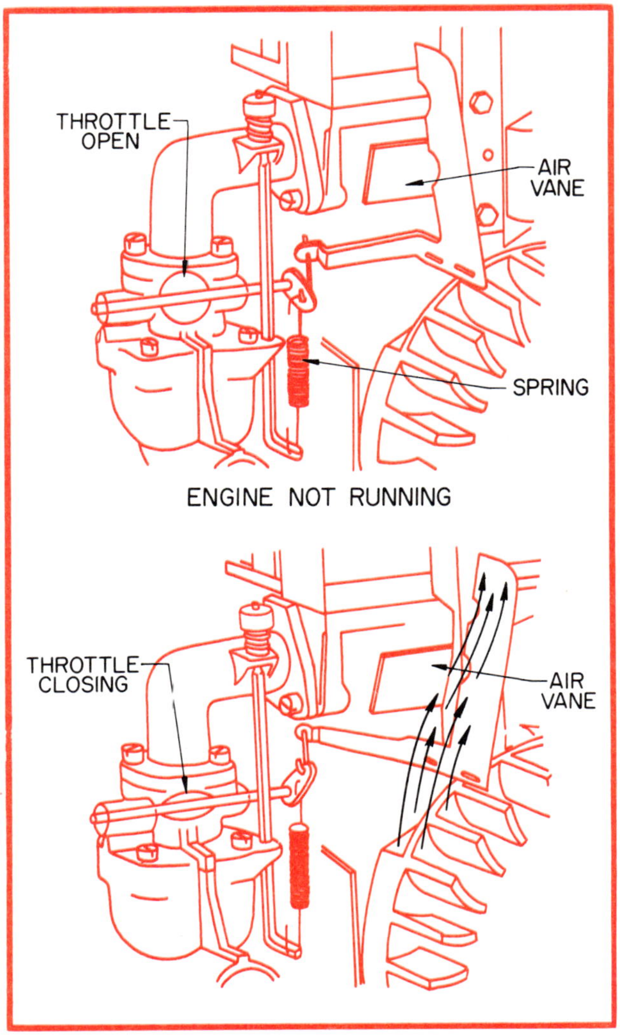

Fig. 5-24. Air vane governor.

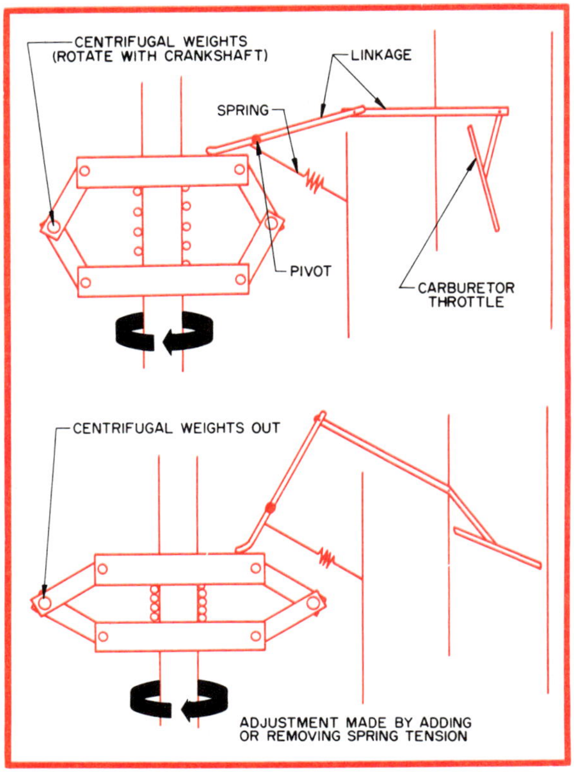

Fig. 5-25. Operation of a centrifugal governor.

throttle which slows down the engine. As the engine slows down, the governor spring pulls the throttle open. Engine speed is increased. The governor spring is connected to a tension control. This lets the operator change or set the engine speed at will.

Mechanical

The mechanical or flyball governor affects engine speed much like the pneumatic governor does. Centrifugal force of weights moving in and out moves the throttle linkage. This controls the speed of the engine, Fig. 5-25.

DISASSEMBLY AND INSPECTION

The following disassembly procedure is general. The exact sequence may need to be changed for certain engines of different manufacturers.

Disassembly and Inspection of the Four-Stroke Cycle Engine

1. Remove the spark plug wire from the spark plug. This prevents accidental starting of the engine.

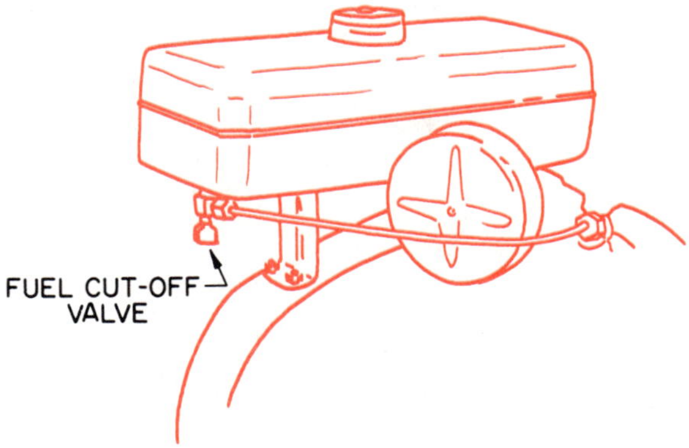

Fig. 5-26. Fuel cutoff valve.

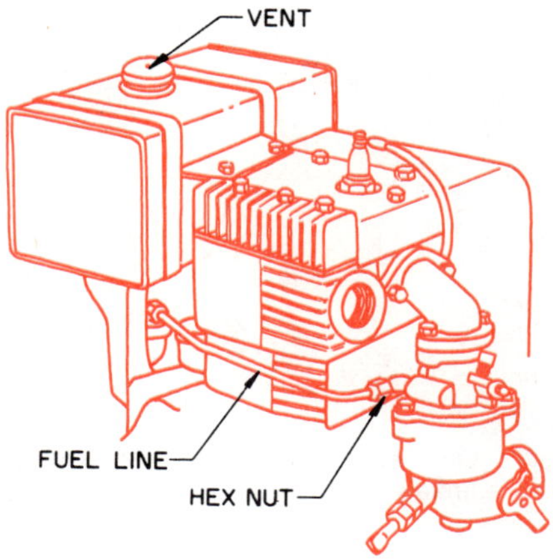

Fig. 5-28. Fuel line.

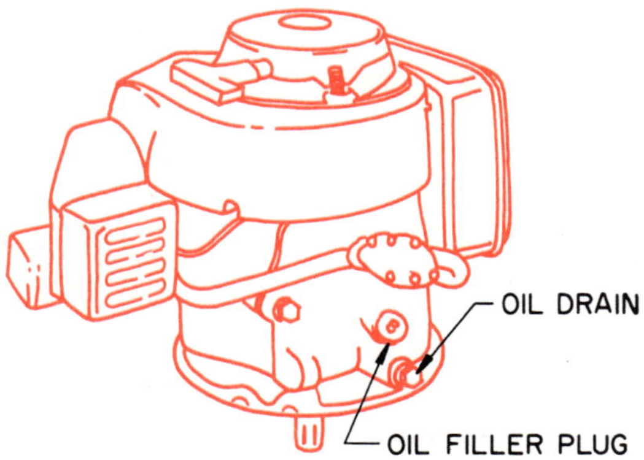

Fig. 5-27. Oil drain plug.

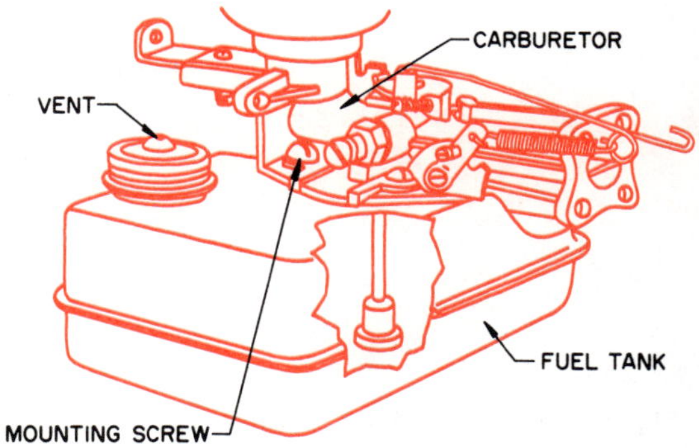

Fig. 5-29. Vacuum feed carburetor.

2. Turn the fuel cutoff valve to the **off** position. Some engines do not have a fuel cutoff valve, Fig. 5-26.
3. Drain the oil from the engine by removing the oil drain plug, Fig. 5-27.
4. Drain the gasoline from the fuel tank. (There is usually a drain plug in the bottom of the gas tank.)

Safety Note

Gasoline should be placed in a safety can. Do not keep it at the workbench or pour it in the sink.

5. Remove the air filter. Check the filter to see if it needs to be cleaned.
6. Drain the gasoline from the fuel line by loosening the lower coupling. Remove the fuel line. Be careful not to twist or break the line. Also, be careful not to "round off" the hex nuts that hold the line in place, Fig. 5-28.
7. Remove the fuel tank. You may need to remove a vacuum feed carburetor before the fuel tank can be removed, Fig. 5-29.
8. Remove the fuel filter and clean it in a suitable cleaner. Allow it to air dry.

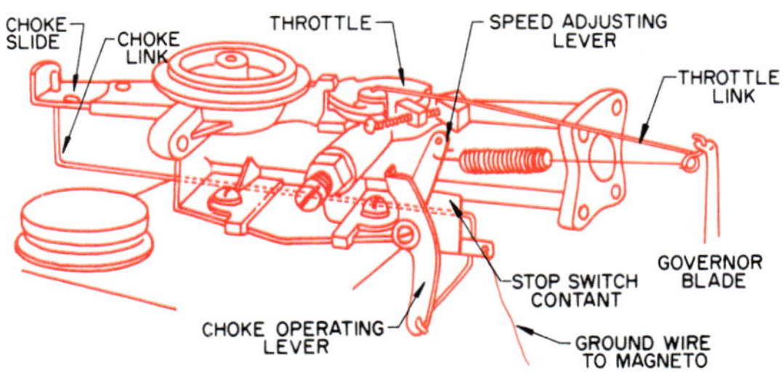

Fig. 5-30. One example of throttle and governor linkage.

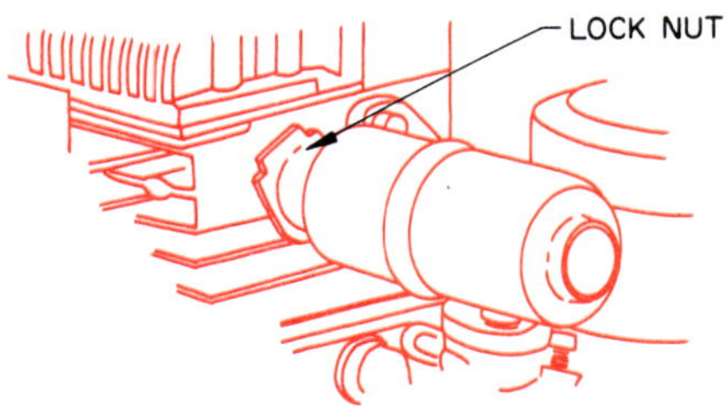

Fig. 5-31. Muffler bolted to engine.

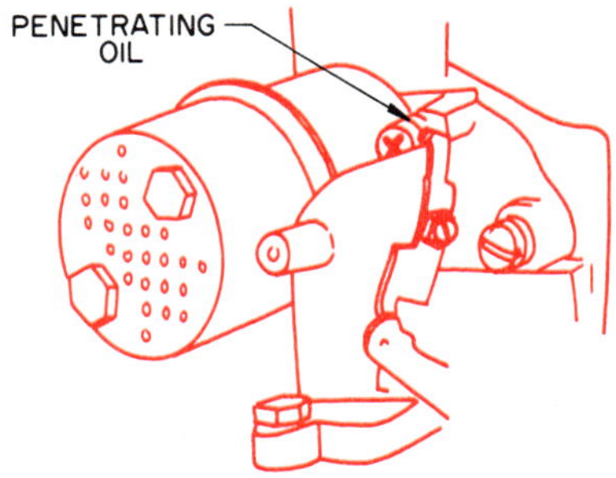

Fig. 5-32. Use penetrating oil on the bolts.

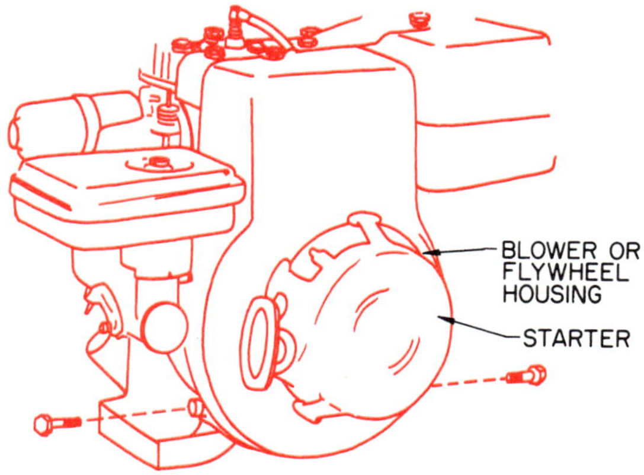

Fig. 5-33. Rope recoil starter.

9. Remove the carburetor. Note the location of the throttle and governor control linkage, Fig. 5-30.
10. Remove the muffler. First loosen the lock nut, Fig. 5-31, if there is one. Place several drops of penetrating oil on the muffler screw threads. Some mufflers are bolted to the engine. Others may be threaded into the exhaust opening of the engine. If the muffler is bolted, place penetrating oil on the bolts, Fig. 5-32.
11. Remove the screws or bolts that hold the blower or flywheel housing to the block. You do not need to remove the recoil starter before removing the blower or flywheel housing, Fig. 5-33.
12. Remove the spark plug with a 13/16 hex deep socket or a special spark plug wrench. Do not remove or replace the spark plug in an aluminum head when the engine is hot. The thread may be damaged.
13. Remove the starter device or recoil ratchet, Fig. 5-33. Hold the flywheel with a flywheel holder and remove the crankshaft nut. Use a special starter clutch wrench on Briggs & Stratton engines, Fig. 5-34.
14. You can remove the flywheel in several ways. One way is to place a knock-off nut on the threaded end of the crankshaft. The nut is then hit sharply with a hammer. The knock-off nut protects the threads and the end of the crankshaft from being damaged,

Fig. 5-35. If you do not have a knock-off nut, place a soft face mallet (rawhide, plastic, or wood) on the end of the crankshaft. Then strike the mallet with a hammer. A special flywheel puller may sometimes be used, Fig. 5-36. A key keeps the flywheel and crankshaft aligned. It may come loose and will have to be put back in place for assembly, Fig. 5-37.

15. Remove the cylinder head bolts, cylinder head, and shield. Make a note of the locations of long and short bolts. When replacing the cylinder head, torque the head bolts. Check the manufacturer's specifications for the correct torque and torquing sequence.

16. Remove the breather assembly, Fig. 5-38. Note the position of the assembly and any connecting tubes. Clean the filter element in the breather assembly, Fig. 5-39.

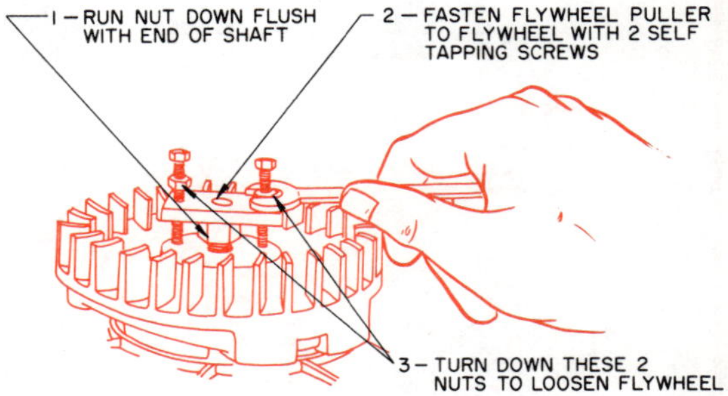

Fig. 5-36. Removing flywheel.

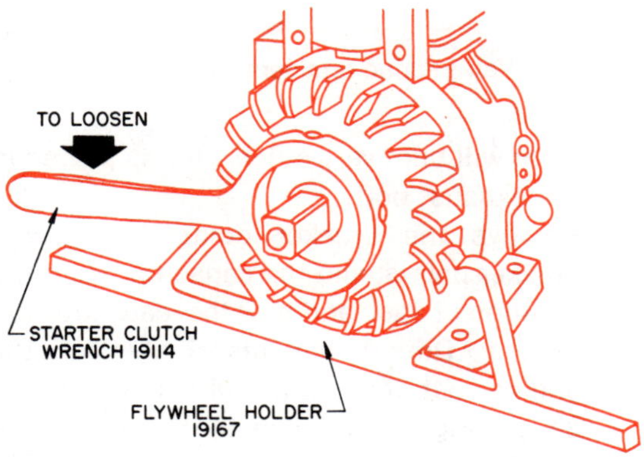

Fig. 5-34. Loosening the flywheel

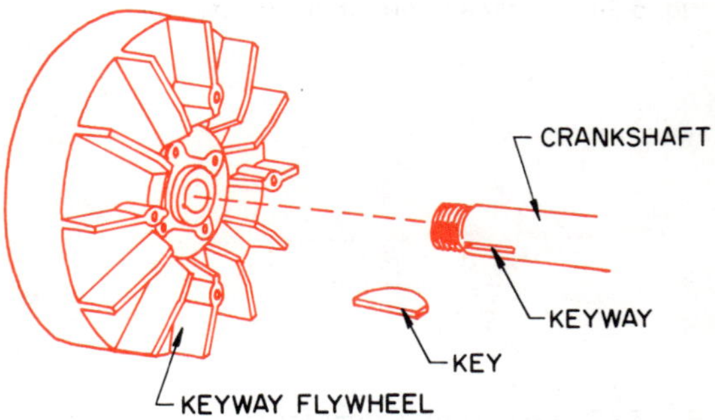

Fig. 5-37. Flywheel keyed to shaft.

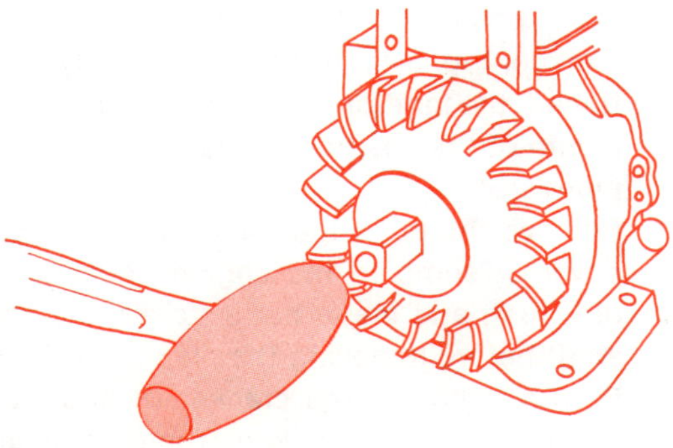
Fig. 5-35. Remove flywheel.

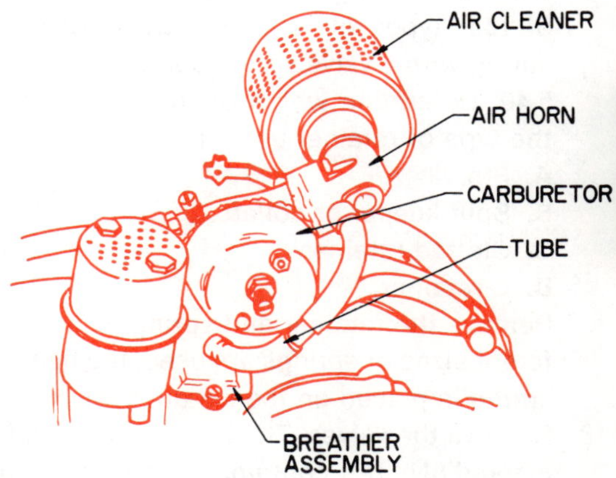

Fig. 5-38. Tied-in crankcase breather assembly.

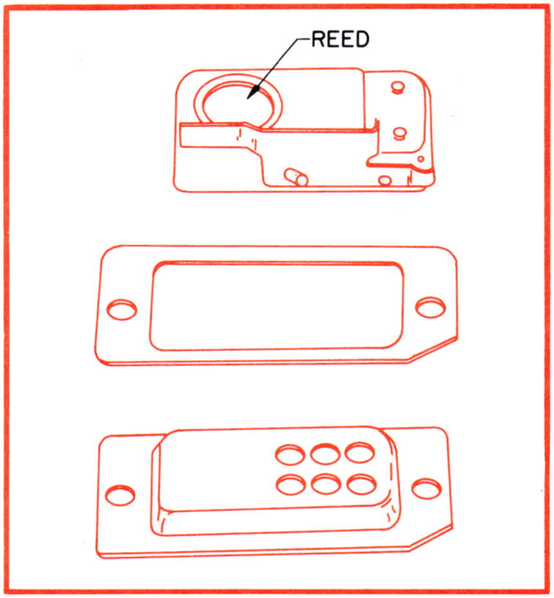

Fig. 5-39. Crankcase breather disassembly.

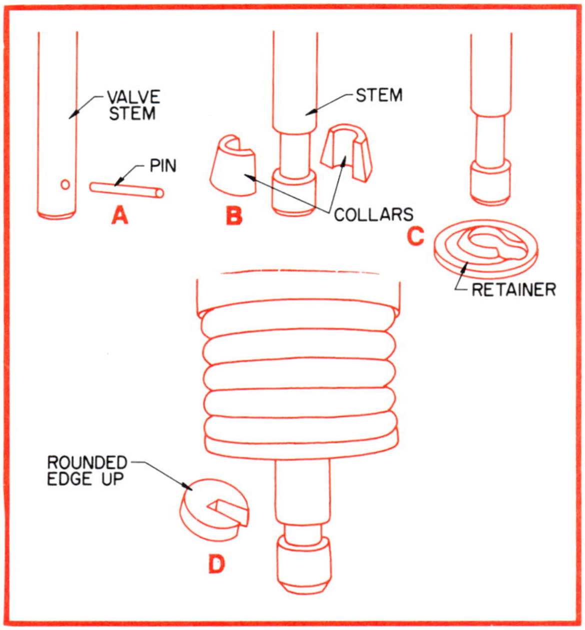

Fig. 5-41. Valve spring retainers.

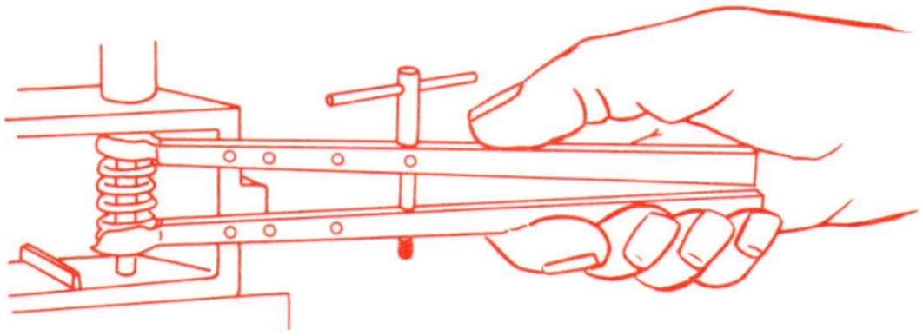

Fig. 5-40. Valve spring compressor.

17. The valve springs are held in place by one of four types of retainers. Compress the spring with a valve spring compressor, Fig. 5-40. Determine which of the following is the type of retainer used, Fig. 5-41:
    A. Pin.
    B. Split keeper or collars.
    C. Slotted retainer.
    D. C-retainer.
    Remove the retainer and spring. If two different sizes of springs are used, the heavier spring is placed on the exhaust valve.
18. Remove the valves. The intake valve will be marked (IN). If it contains no markings, the exhaust valve will be marked (X). The intake valve will burn very quickly if it is placed in the exhaust port.
19. Remove the breaker point dust cover, breaker points, and the capacitor. Note the position of each and how the leads are connected. During the reassembly of the engine, set the points with a flat feeler gauge, Fig. 5-42. Clean the points with a piece of paper. Inspect the breaker points for excessive pitting, Fig. 5-43.
20. Remove the breaker box assembly. It may be a part of the magnetic ignition coil. The exact position of the assembly should be marked, Fig. 5-44. If it is not properly positioned, the ignition timing will be off. Make a note of the position of the breaker can or plunger.
21. Remove the bolts that hold the crankcase cover or sump to the engine. Remove all dirt, rust, and metal burrs from the crankshaft before removing the plate, Fig. 5-45.
22. Before removing the crankshaft and camshaft, locate the crankshaft and camshaft timing marks, Fig. 5-46. If there are no tim-

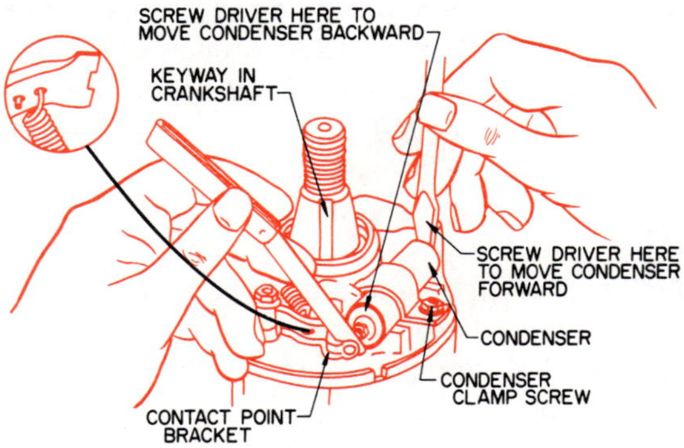

Fig. 5-42. Adjust breaker point gap.

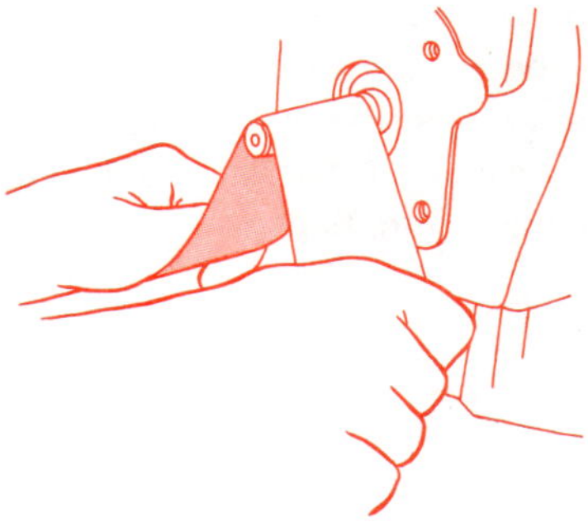

Fig. 5-45. Removing rust and burrs.

Fig. 5-43. Excessive pitting of contact points.

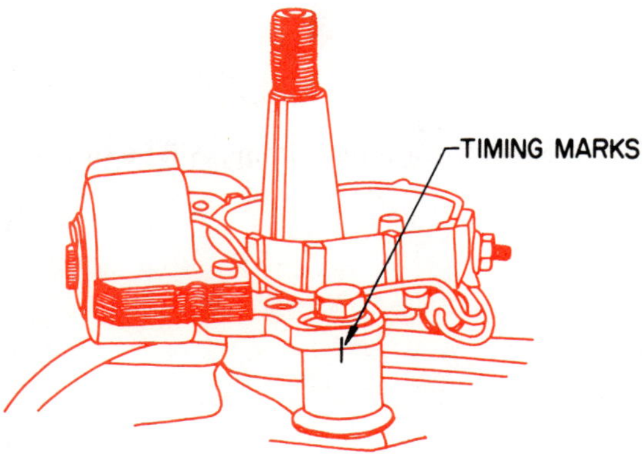

Fig. 5-44. Typical timing marks on four-cycle engine.

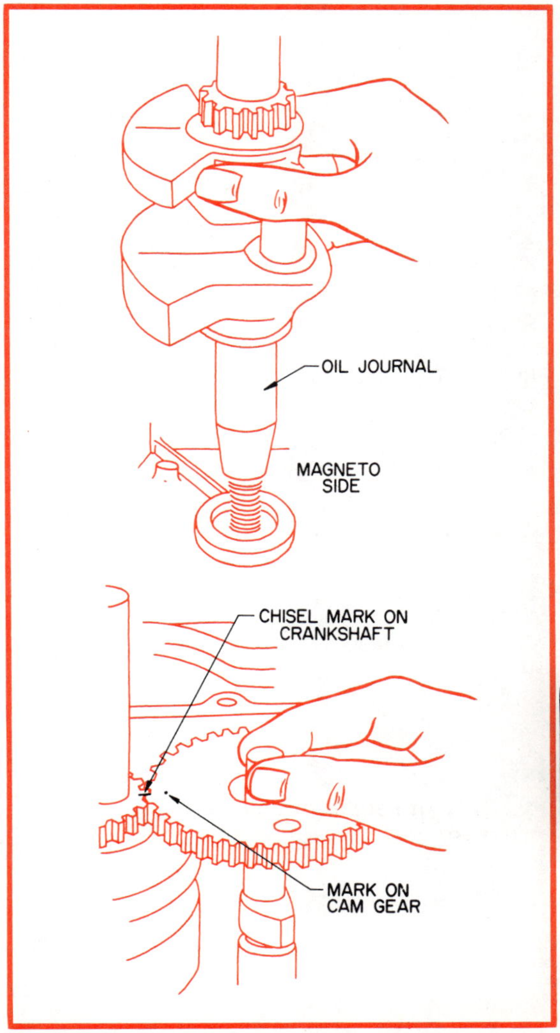

Fig. 5-46. Aligning timing marks.

ing marks, use a center punch and mark the position of the camshaft and crankshaft. Note the position of the oil pump or slinger if one is used. Also, remove the valve lifters or tappets, Fig. 5-47.

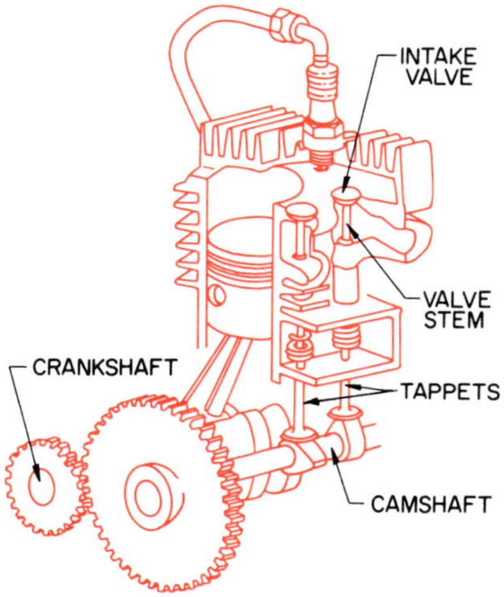

Fig. 5-47. Tappets.

23. Remove the connecting rod bolts and lock, Fig. 5-48. Use a socket wrench or box end wrench to prevent rounding of the bolt heads. Note the position of the following elements, Fig. 5-49.
    a. The assembly marks on the connecting rod.
    b. The oil hole in the connecting rod.
    c. The oil dipper, if one is used.
    d. The cam gear side of the connecting rod.

    The connecting rod bolts should be torqued when the engine is reassembled. Check the manufacturer's specifications for the exact torque.
24. Push the connecting rod and piston out through the top of the cylinder. Inspect the rod bearing, piston rings, piston, and cylinder for excessive wear. Use a ring compressor when placing the piston back in the cylinder, Fig. 5-50.
25. Pull the crankshaft out slowly, twisting it slightly if necessary. Check the bearing surface for wear.

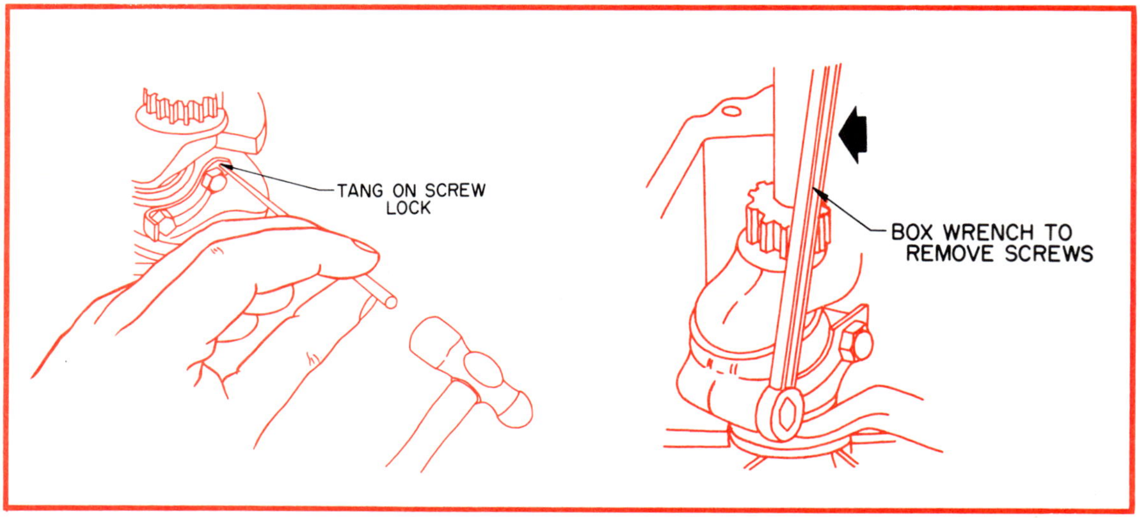

Fig. 5-48. Bending rod lock.

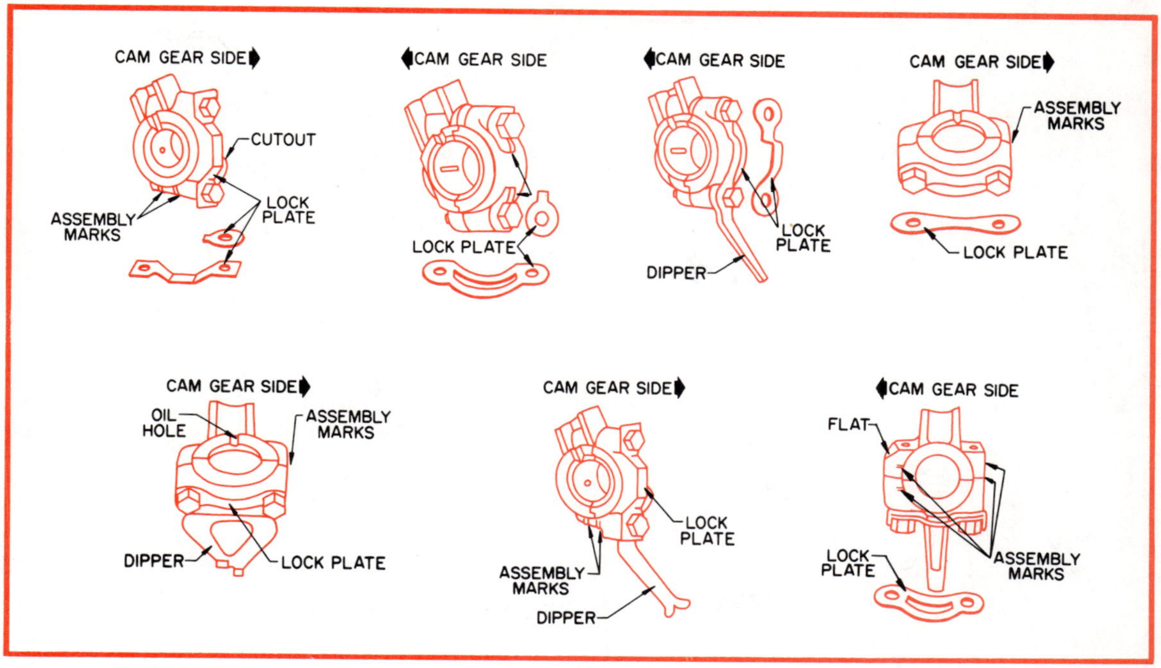

Fig. 5-49. Connecting rod installation.

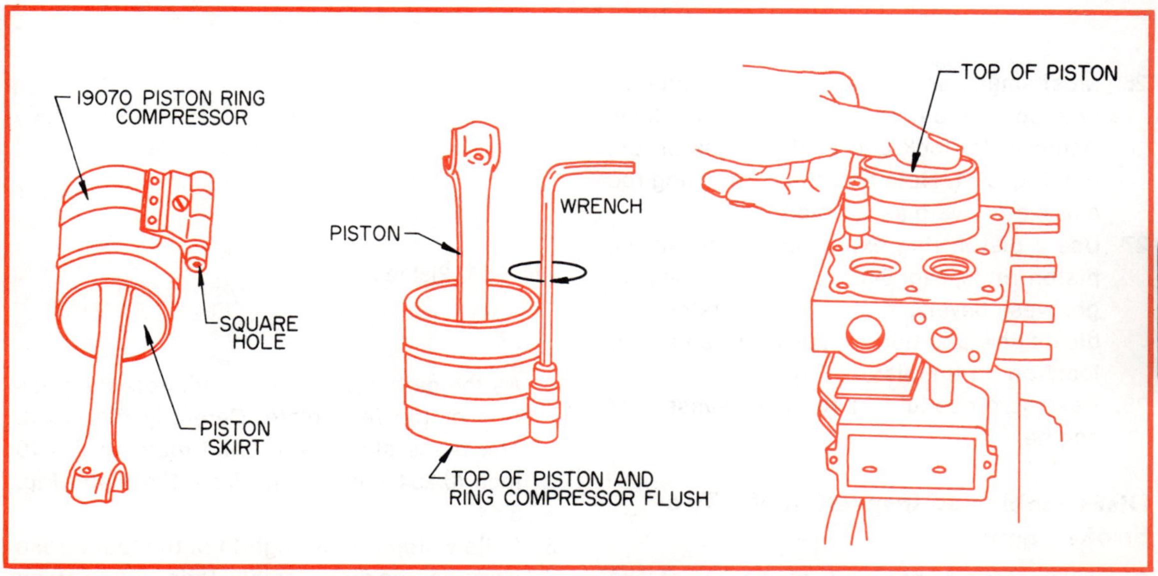

Fig. 5-50. Install piston assembly.

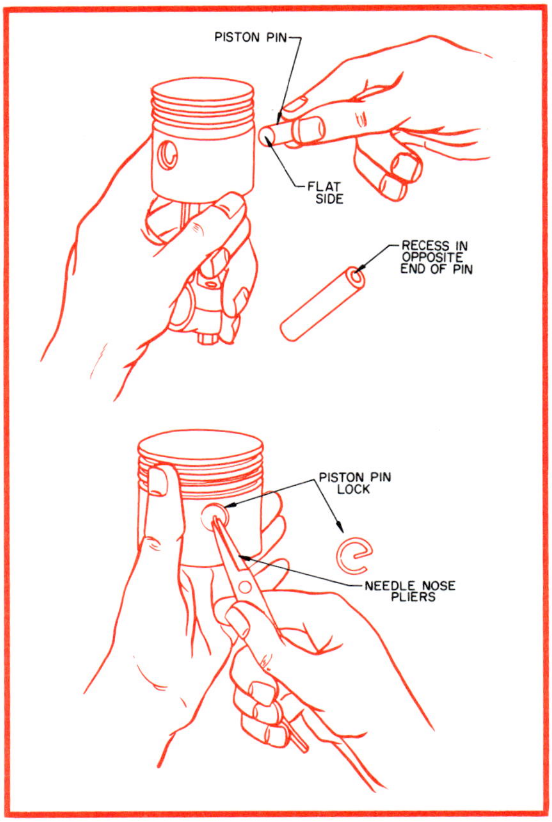

Fig. 5-51. Removing piston pin.

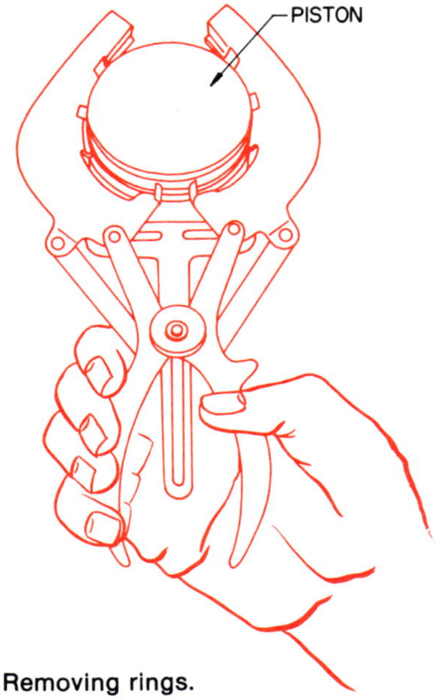

Fig. 5-52. Removing rings.

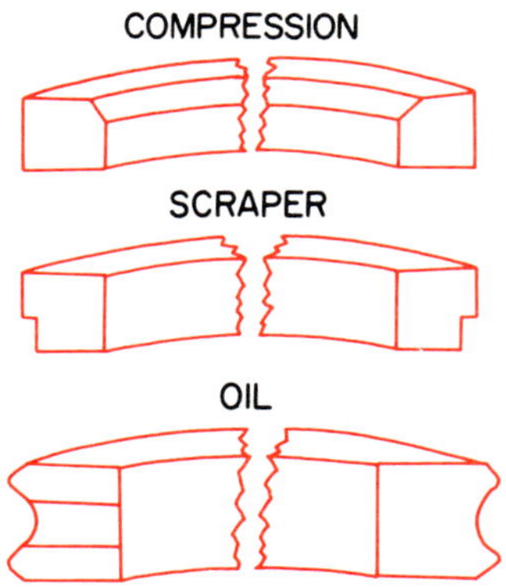

Fig. 5-53. Piston rings.

26. Most small gasoline engines use a pin lock or snap ring to hold the piston pin in place. Remove the lock and push the piston pin out, Fig. 5-51. Note how the connecting rod and piston were assembled.
27. Use a piston ring expander to remove the piston rings, Fig. 5-52. Check the rings for grooves, bevels, etc. These markings indicate the position of the rings and their location on the piston, Fig. 5-53.
28. Reverse procedures 1 to 27 to reassemble engine.

Disassembly and Inspection of a Two-Stroke Engine

1. Follow steps 1 through 9 of the four-stroke cycle engine disassembly.
2. As the carburetor is removed, note the position of the reed plate. Carefully remove it. The reeds should not bend more than .010 inch (0.254 millimeters) from the plate, Fig. 5-54.
3. Follow steps 10 through 14 of the four-stroke cycle engine disassembly, then skip to steps 19 and 20.

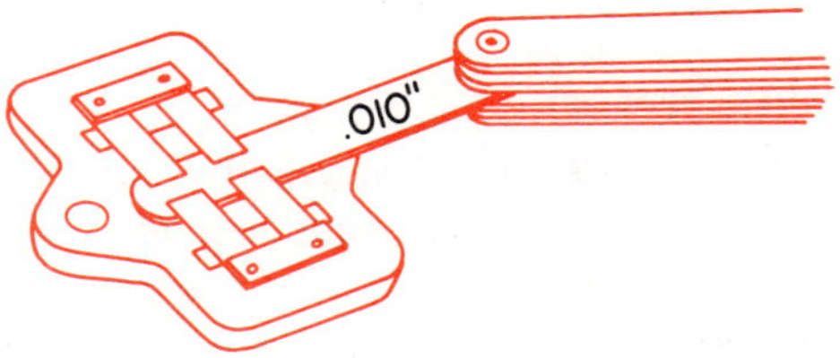

U.S. Customary	SI Metric
.010 in.	0.254 mm

Fig. 5-54. Carburetor reeds.

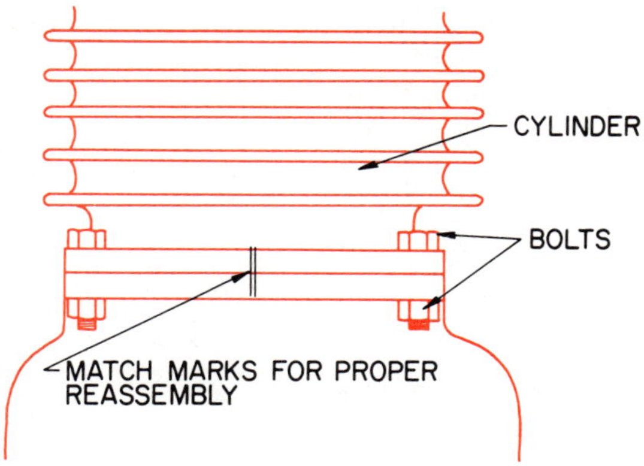

Fig. 5-55. Alignment of crankcase and cylinder block.

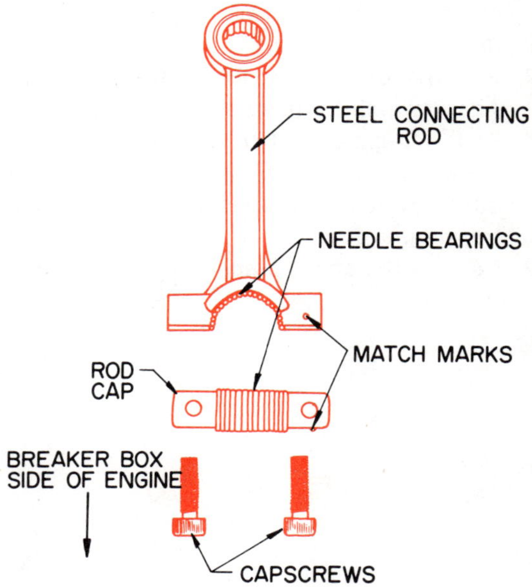

Fig. 5-56. Steel connecting rod.

4. Many two-stroke cycle engines do not have removable cylinder heads. Mark the cylinder block position on the crankcase. Remove the bolts that hold the cylinder block to the crankcase. Pull the cylinder block off of the piston assembly, Fig. 5-55.
5. Remove the screws from the connecting rod. Note the position of the following elements, Fig. 5-56.
   a. The connecting rod cap and connecting rod (match marks).
   b. The side of the connecting rod that faces the breaker box assembly.
   c. The needle bearings that may be on the connecting rod.
6. Remove the mounting screws in the crankcase cover, Fig. 5-57, or separate the

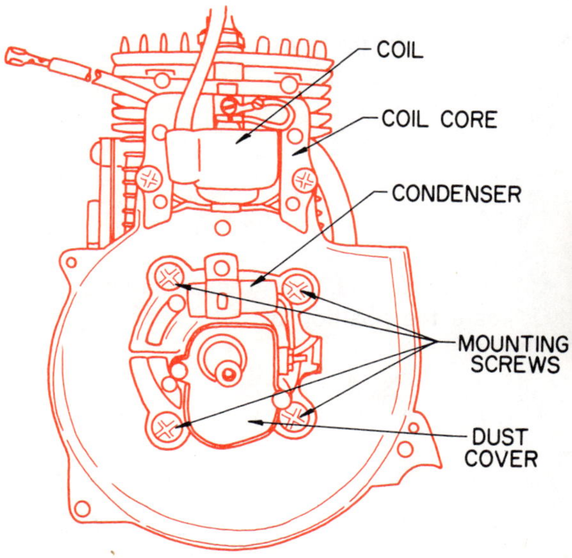

Fig. 5-57. Remove mounting screws.

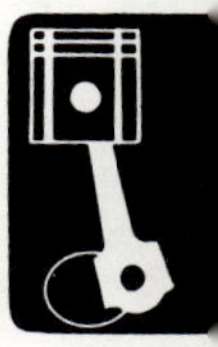

crankcase, Fig. 5-58. Watch for needle bearings.

7. Pull the crankshaft out of the crankcase. Again, watch for needle bearings.
8. Follow steps 26 and 27 of the four-stroke cycle engine disassembly. Note that the two-stroke cycle engine does not have an oil control ring.
9. Reverse the above precedure to reassemble the engine.

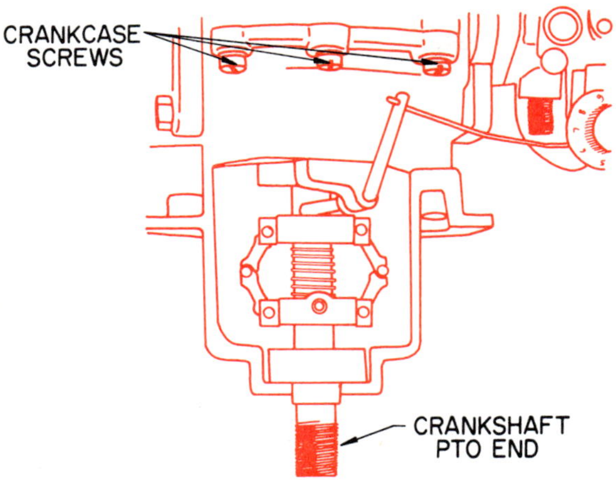

Fig. 5-58. Separate the crankcase.

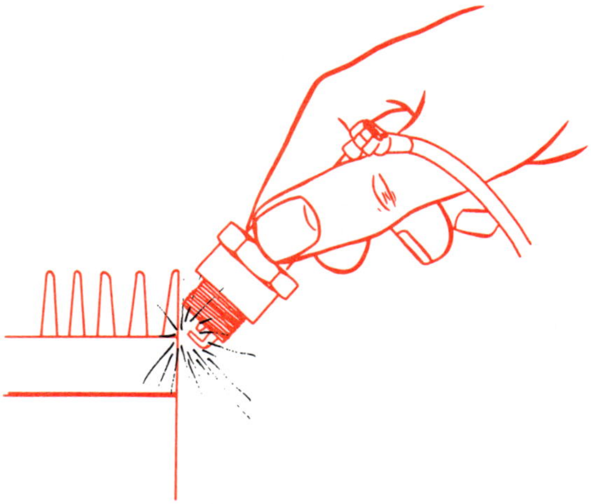

Fig. 5-59. Checking spark plug.

TROUBLESHOOTING

Ignition System

Check for spark. Remove the spark plug and hold it against the cylinder block. Do not touch the top of the spark plug. You may get an electric shock. Pull the starter rope. A spark

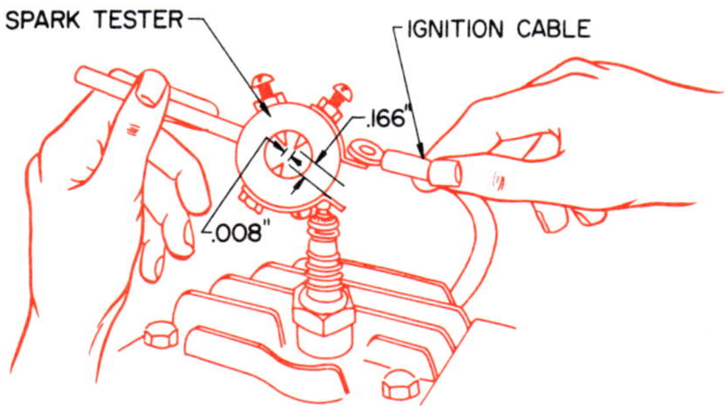

U.S. Customary	SI Metric
.008 in.	2.0320 mm
.166 in.	4.2164 mm

Fig. 5-60. Testing spark plug.

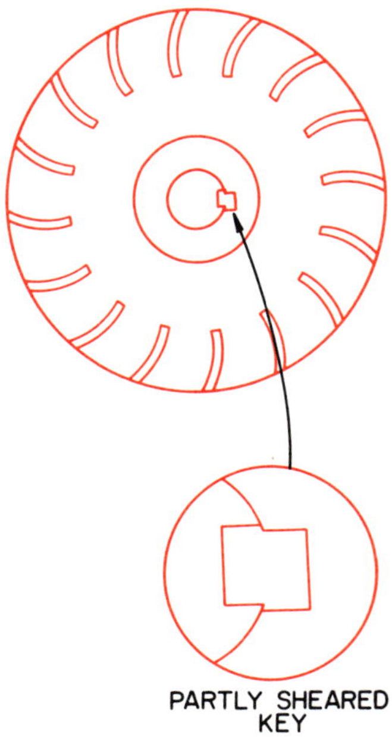

Fig. 5-61. Partly sheared key.

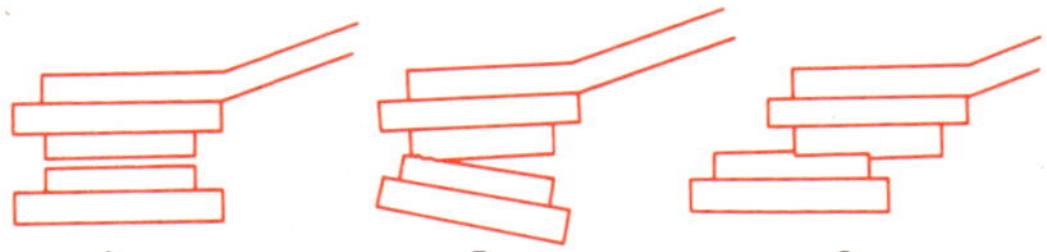

Fig. 5-62. Breaker point alignment.

should jump the air gap of the spark plug, Fig. 5-59. If no spark occurs, check the following items in order until the problem is corrected.

1. Try a new spark plug or check the old one with a spark plug tester, Fig. 5-60.
2. Check the spark plug air gap.
3. Check for sheared flywheel key, Fig. 5-61.
4. Check the breaker point gap.
5. Clean the breaker points.
6. Check for pitted or misaligned breaker points, Fig. 5-62.
7. Check for loose connection or shorted wiring.
8. Check the flywheel-to-laminated iron core air gap.
9. Check for a bad on-off switch or "kill" switch.
10. Check the capacitor.
11. Check the magnetic coil.
12. Check the flywheel magnets by using a good flywheel.
13. Check for correct ignition timing. Follow the manufacturer's timing procedure.

Compression

There are two methods of checking the compression of the engine. The first method will determine if the compression is high enough to start the engine. The second method will give the exact pressure reading inside the engine.

1. Pull the starter rope and quickly release it. If there is a strong snap-back or bounce-back, the engine should start. On engines equipped with "Easy-Spin" starting systems, the engine may have to be turned counterclockwise.
2. Remove the spark plug and ground the spark plug wire. Place a compression gauge in the spark plug hole. Turn the engine over with the starter rope. The compression reading should be 60 psi (413.7 kilopascals) or above, Fig. 5-63.

If the compression check is below the recommended reading, look for the following causes:

a. Bad threads in spark plug hole.
b. Loose cylinder head bolts.
c. Bad head gasket.
d. Burnt valves or valve seats.
e. Insufficient valve stem to valve tappet clearance.
f. Sticking valves.
g. Warped cylinder head.
h. Bad rings.
i. Broken connecting rod.

Fuel System

1. Check the gas tank for clean fuel.
2. Check for closed cutoff valve in the gas line.
3. Check for plugged fuel filters or fuel line.
4. Check for flooding.
   a. Overchoking.
   b. Excessively "rich" fuel mixture. (Carburetor needs adjusting.)
5. Check to see if fuel pump is working.
6. Check for dry spark plug. If spark plug is dry, remove the spark plug and put about 2

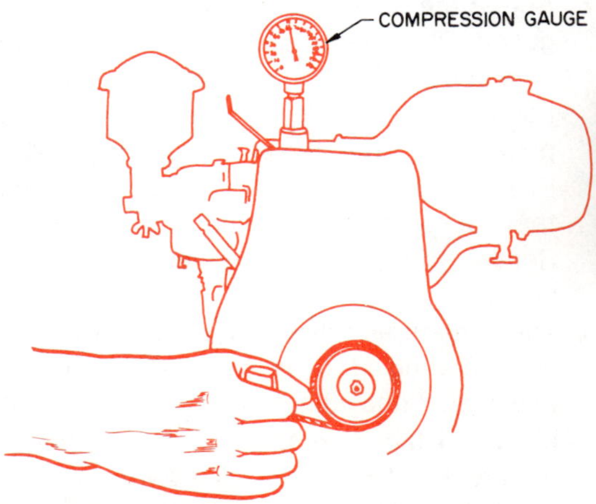

Fig. 5-63. Checking compression.

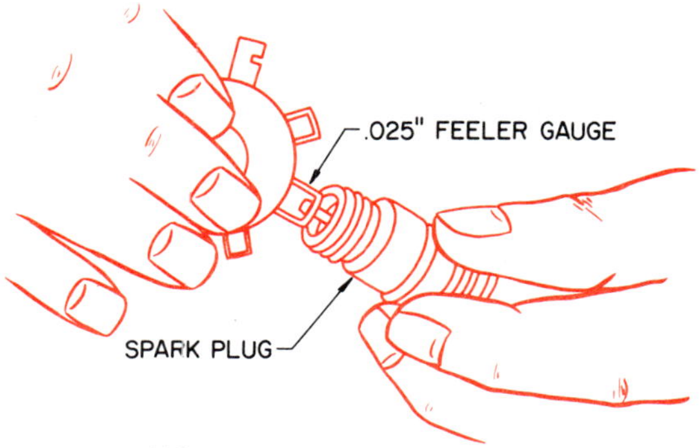

U.S. Customary	SI Metric
.025 in.	0.635 mm

Fig. 5-64. Adjusting spark plug gap.

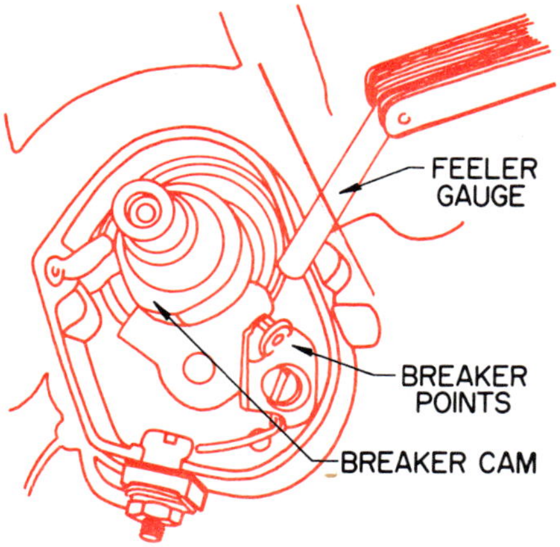

Fig. 5-65. Gap points.

tablespoons (30 milliliters) of fuel in the spark plug hole. Replace the spark plug and try to start the engine. If the engine fires a few times and quits, check for the following causes:

a. Loose carburetor.
b. Plugged screens or jets.
c. Fuel control system not operating or gummy (float system or diaphragm).

Ignition System

1. Adjust the spark plug gap to the manufacturer's specifications. Use a wire feeler gauge, Fig. 5-64.
2. Adjust the breaker points with a flat feeler gauge. Place the high portion of the cam lobe under the rubbing block of the points. Loosen the hold-down screw. Adjust the points with the feeler gauge, Fig. 5-65. Place a piece of clean white paper between the closed points and pull it out, Fig. 5-66. Repeat this until the paper comes out clean.
3. The armature-to-flywheel air gap may be nonadjusting if it is located under the flywheel. If it is adjustable, place a piece of electrical tape across the poles of the armature, Fig. 5-67. Place the flywheel on the crankshaft. Slowly rotate the magnet over

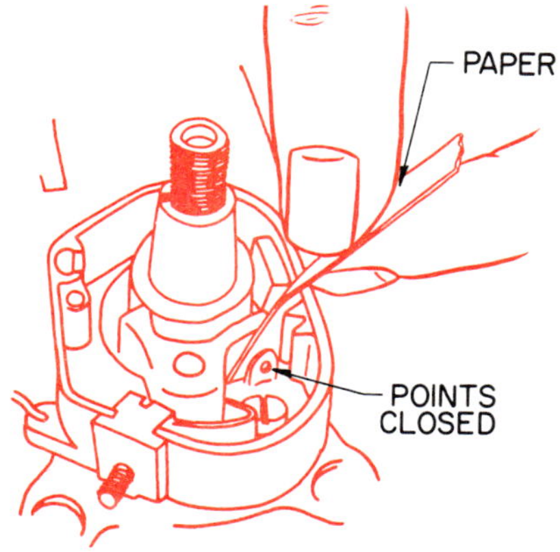

Fig. 5-66. Clean points.

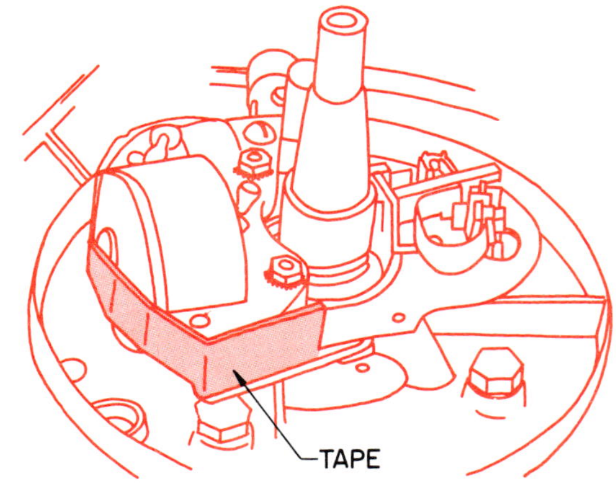

Fig. 5-67. Checking gap between flywheel and armature.

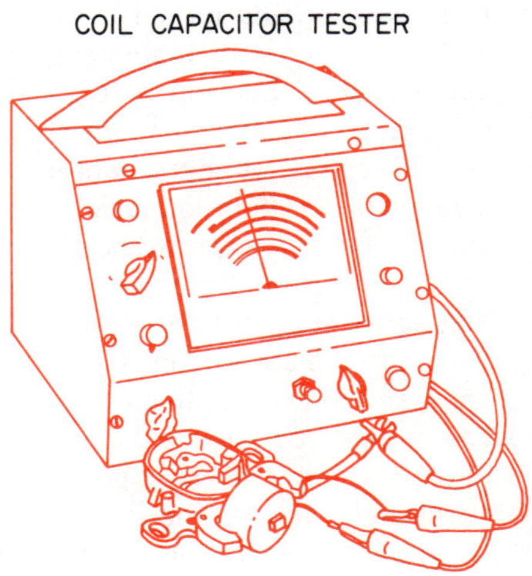

Fig. 5-69. Check parts carefully.

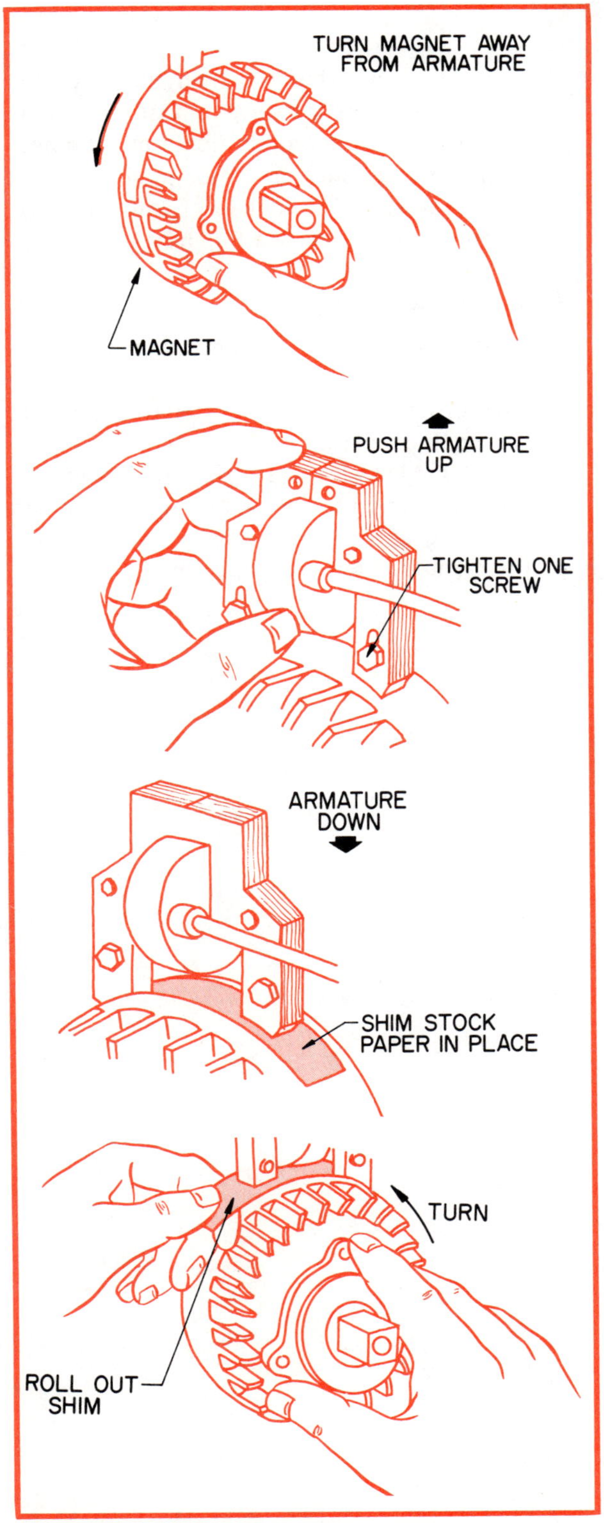

Fig. 5-68. Adjusting armature air gap.

the armature. If the tape is scuffed, the air gap is too small. If the magnet will not scuff a second layer of tape, the air gap is too large. Tape should be about .008 inch thick (0.2032 millimeters). When the armature is on the outside of the flywheel, several layers of paper the proper thickness can be placed over the flywheel magnet and under the armature, Fig. 5-68. Loosen the armature screws. The magnets will pull the armature down. Tighten the armature bolts and remove the paper.

4. Test the capacitor on a capacitor tester. Fig. 5-69. Or replace it with a capacitor that is known to be good.
5. Test the coil on a coil tester or replace it with a good coil.
6. The timing of a small gasoline engine very seldom gets out of adjustment. Study the manufacturer's service manual when you attempt to set the ignition timing.

Fuel System

1. Remove fuel filters and clean them with a suitable solvent.
2. There are several adjustments that may be made on the carburetor.

a. Idle speed should be set between 1200 and 1500 rpm with the idle speed adjusting screw. See Fig. 5-70. Check the manufacturer's specifications.
b. The low speed jet adjustment should be made with the throttle closed (idle rpm). Turn the idle valve approximately one-half turn open. Start the engine. Screw the idle valve out until the engine falters. Then turn the adjustment screw half way between the two positions.
c. The high speed jet is adjusted with the throttle about one-half open. Turn the needle valve **in** until the engine starts to cut out. Then turn the needle valve **out** until the engine starts to cut out. Turn the needle valve halfway between the two positions tested. Close the throttle and then quickly open it. The engine should not stall or hesitate. If it does, turn the high speed needle valve **out** a small amount. Accelerate the engine again. Repeat the above procedure until the engine accelerates smoothly.
d. If the above adjustments do not solve any carburetor problems, disassemble and clean the carburetor in a carburetor cleaning solvent. Check all needle valves for damage, Fig. 5-71. Adjust the float level if there is one. This is done by forming the tab that holds the valve in position. Many small gasoline engine floats may be set by turning the carburetor upside down and setting the float level with the carburetor housing. See Fig. 5-72.

Governor Systems

1. Check the governor system for misaligned linkage or binding parts.
2. Most governors may be adjusted by changing the position of the linkage. Or adjustment can be made by increasing or decreasing the governor spring tensions. For instruction on adjusting specific types of governors, see the manufacturer's specifications.

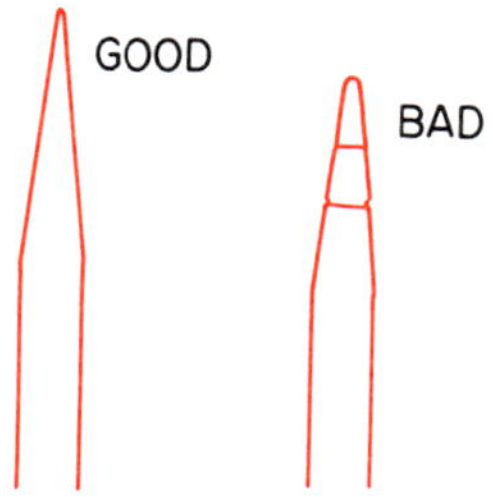

Fig. 5-71. Damage or wear on needle valves.

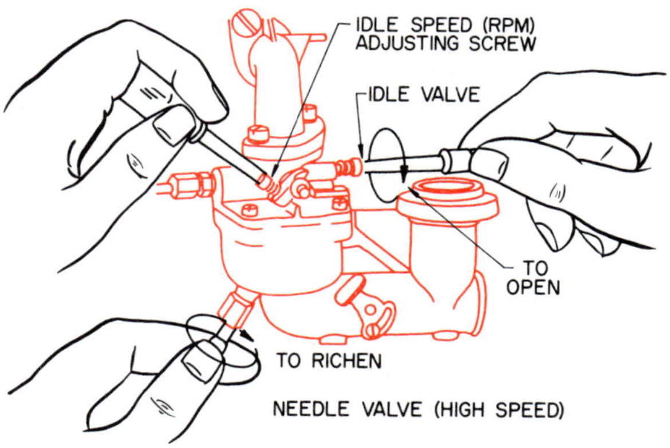

Fig. 5-70. Final speed adjustment.

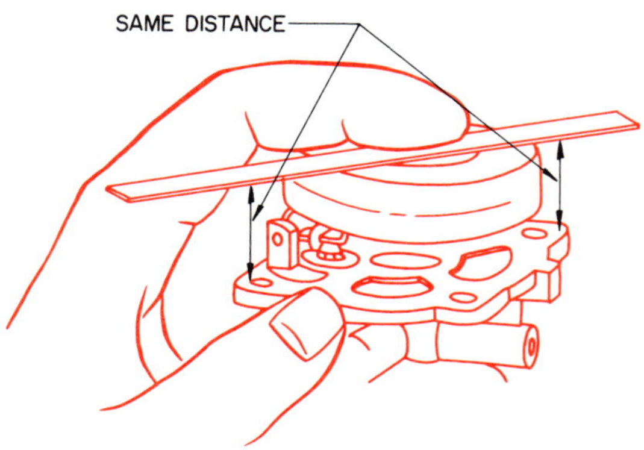

Fig. 5-72. Checking float level.

Lubrication System

Many small gasoline engine manufacturers suggest that you change the oil every 10 to 20 hours of operation. Drain the oil. Replace it with clean oil that has a service rating as specified in the manual.

Cooling System

The cooling system should be cleaned whenever dirt and grass collect or when the engine is disassembled. Clean the following parts:

1. Screen over the flywheel.
2. Radiating fins of the head and cylinder block.
3. Flywheel vanes.
4. Blower housing.

Valves and Valve Seats

1. When an engine has a loss of compression, the valves and valve seats may be burned. Remove the valves and renew the valve face. Use either a valve grinder or cutter.
2. Most valves should be renewed to a 45 degree angle, although some valves are cut at a 30 degree angle. The margin of the valve should not be less then 1/64″ (0.397 millimeters). See Fig. 5-73.

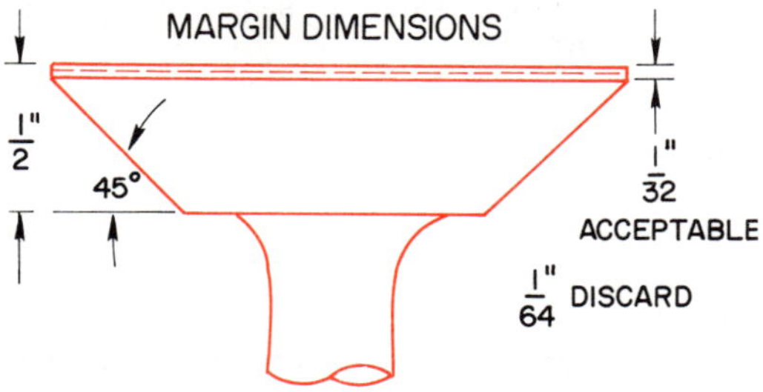

U.S. Customary	SI Metric
1/64 in.	0.39 mm
1/32 in.	0.79 mm
1/2 in.	12.70 mm

Fig. 5-73. Valve margin dimensions.

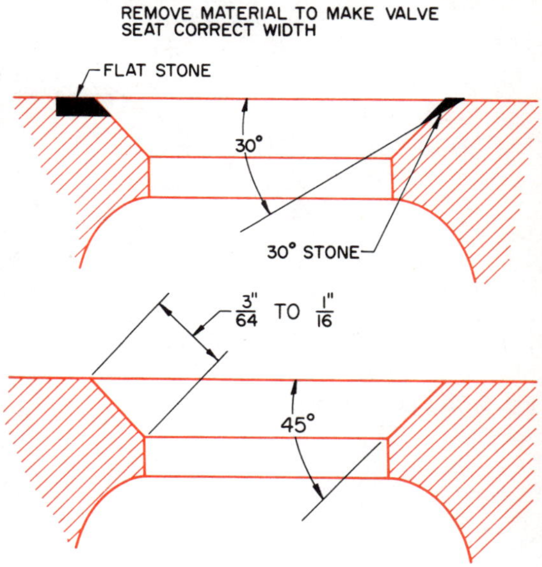

U.S. Customary	SI Metric
3/64 in.	1.19 mm
1/16 in.	1.58 mm

Fig. 5-74. Adjusting valve seats.

3. The valve seats should also be renewed. This can be done by using a valve seat grinder or cutter. Most valve seats are cut at a 45 degree angle, but some are 30 degrees, Fig. 5-74. Suppose the valve does not seat in the center of the valve face. The valve seat may then have to be cut on the top surface with a different angle stone or cutter. Use a flat or 30 degree stone on 45 degree valve seats. Use a flat or 15 degree stone on 30 degree valve seats to obtain the proper valve seat width. See Fig. 5-74.
4. After the valves and valve seats have been ground, you need to adjust the valve stem-to-tappet clearance. This is done by grinding material from the end of the valve stem. The clearance is measured with a feeler gauge, Fig. 5-75. Grind carefully. The stem should remain square. Be careful not to remove too much material.

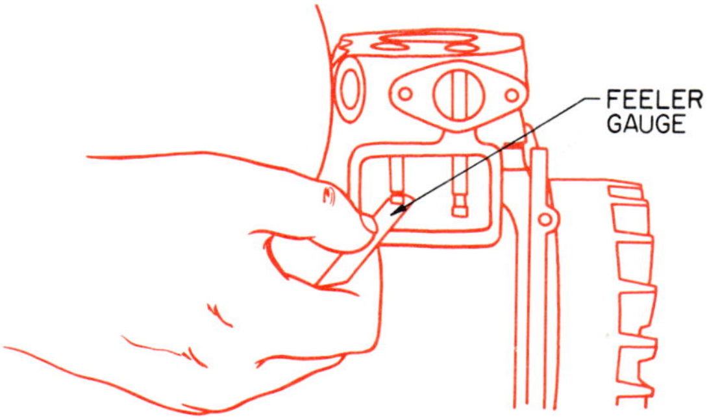

Fig. 5-75. Checking clearance.

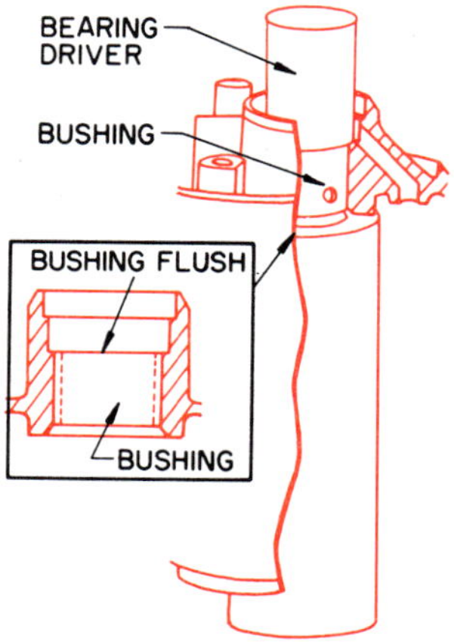

Fig. 5-77. Pressing in new bushing.

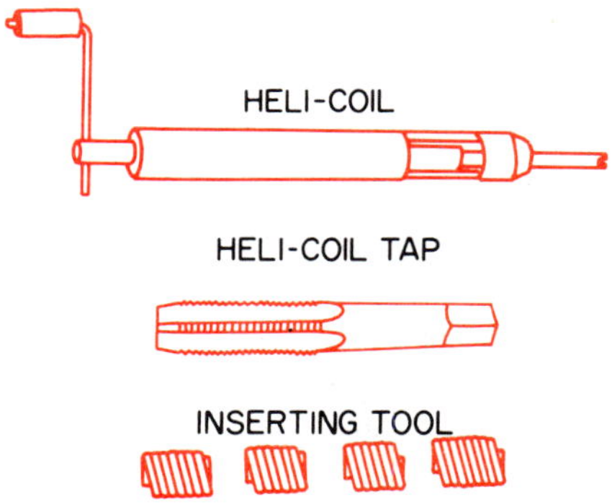

Fig. 5-76. Repairing threads with heli-coil.

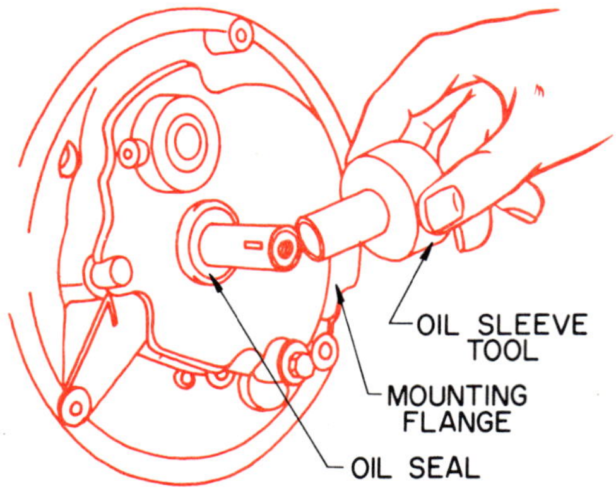

Fig. 5-78. Mounting flange oil seal with an oil seal valve tool.

Cylinder Heads

1. Loss of compression may result from a warped cylinder head. To check this, remove the head and place it on a very flat surface. If a .005″ (0.1270 mm) feeler gauge will slip under the head at any point, renew the gasket surface. Place a piece of abrasive paper on a flat surface. Move the head in a figure eight pattern on the paper until the head gasket surface is flat.
2. Clean the fins and check for broken fins. Replace the cylinder head if any fins are broken. If the spark plug hole threads are damaged, renew the threads with a heli-coil. See Fig. 5-76.

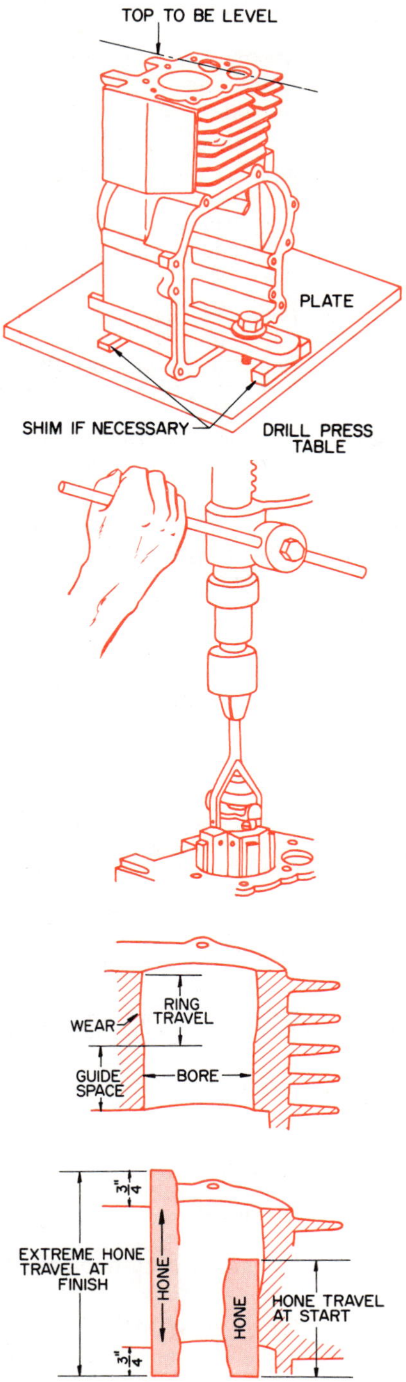

U.S. Customary	SI Metric
3/4 in.	19.05 mm

Fig. 5-79. Honing cylinder.

Cylinder Block

Check the cylinder block for damaged fins, bearings, oil seals, and threaded holes. If any damage is found, repair or replace the cylinder block.

1. Damaged threaded holes may often be repaired by using heli-coils. The holes are drilled oversize. They are threaded with a heli-coil tap, and the heli-coil is inserted.
2. You may be able to remove damaged bearings and press in new bearings, Fig. 5-77. Sometimes the block must be replaced.
3. Whenever an engine is disassembled, the oil seals should be replaced. Pull the oil seal out. Insert a new seal with an oil sleeve tool, Fig. 5-78.
4. If the piston assembly is removed from the cylinder, or if new rings are to be installed, the cylinder should be honed, Fig. 5-79. Use a medium grit stone and break the glaze on the cylinder. The hone pattern should be a cross hatch pattern. Fig. 5-80. Achieve this by moving the hone up and down rapidly. Clean the cylinder walls with clean oil and rags. Wipe the cylinder until the oil-soaked rags come out clean.

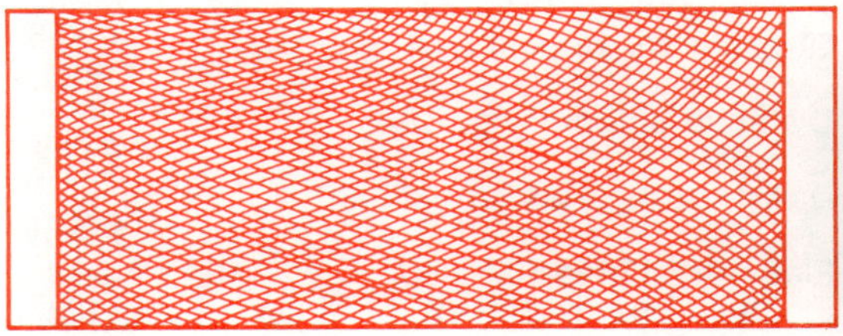

Fig. 5-80. Cross-hatch pattern.

INDEX